A Meaning in the Universe

2024

Copyright © 2024 Henri Gillet
All rights reserved.

Contents

FOREWORD 5
INTRODUCTION 8
CHAPTER I. A psychology of spirit confronted with materialism 17
    A)   Psychology wants to assert itself as a science............................18
    B)   Frankl develops Existential Analysis 20
    C)   Materialist determinism and spiritualist free will 32
CHAPTER II. Scientific materialism is questioned......................................35
    A)   From the immensely large: Cosmology. 37
    B)   To the infinitely small: Quantum physics. 45
        The mystery of measurement in quantum physics..........................48
        Quantum decoherence. 54
        Non-separability: entanglement 55
        The principle of indeterminacy 57
        Quantum time and classical time...57
    C)   Determinism and indeterminability 59
    D)   Matter and reality 62
    E)   Independent reality and empirical reality 63
    F)   Realities and Consciousness 68
    G)   Mathematics and Consciousness 72
    H)   Life sciences are also questioned 77
    I)   Science and spirituality..84
CHAPTER III. Existential Analysis and the Rehabilitation of the Spirit........89

A)	Emergence of the philosophy and psychology of freedom and responsibility.	92
B)	An Existential Analysis enriched with contemporary reflections	99
C)	Existential Analysis elevates the individual into a person	105
D)	Existential analysis and humanization of society	110
E)	Re-enchanting Man and the world	116

CONCLUSION 124

Referenced Authors 133

Bibliographic references 139

Warning: *This book was translated from French and may contain translation errors. The kind reader is invited to contact the person responsible for translation and leave an email with all necessary corrections at gillet.joel@gmail.com. Thank you for your kindness, it makes all the difference.*

FOREWORD

Karl Marx famously said: "It is not a question of understanding the World, it is a question of changing it". I admit that I am among those who always wondered if it was not more relevant to try to understand it before wanting to change it.

I have always been interested in the descriptions and explanations of the world offered by science, from the most rigorous such as cosmology or physics to those aspiring to be so, such as sociology and psychology. I doubtless hoped to satisfy a curiosity that seemed natural to me, but also perhaps to find there what I was looking for, a kind of coherence, significance, meaning. And in fact, I was rather sorry for this world which, more and more, proved to be devoid of it, to the point of suggesting its self-destruction.

When I was younger, my studies, mainly scientific, had interested me in the admirable rigor of mathematics and the sober elegance of physics. But the deterministic materialism that was its foundation, and explained the present and the future exclusively by chance encounters in the past, between matter and forces arising from nothingness, seemed to me intellectually brilliant but philosophically insufficient. It seemed more like a postulate that allowed science to build itself against religion.

In parallel, a certain number of human sciences, as well as neurosciences have depicted man as the exclusive product of his past and his environment, denying him any effective freedom and any real responsibility. Considering others, and myself, in this way seemed equally distressing to me.

Our society attributes a specific dignity to every human being, which can logically only be based on a specific human dimension. So, to observe that science, on which this society is based, contests this specific dimension for man, seemed paradoxical to me.

After the age of fifty, I discovered Viktor Frankl and Existential Analysis. The vision of the man that he described there, that of a free and responsible being in search of meaning, seemed to me of an obvious clarity, but I was frustrated not to be able to establish it on a solid objective basis. Indeed, science denied Frankl the spiritual dimension of man, at the source of this need for meaning.

In recent years, I have been interested in new scientific reflections, both in physics and in psychology, which question this interpretation of the strictly determinist, materialist and reductionist world. The most recent works, in the fields of astronomy, physics, paleontology or psychology, now give a much more nuanced opinion on this absolute determinism from the past. They come to evoke as only possible explanation of a certain number of phenomena, the effect of a project in progress. The implementation of a future is therefore also to be taken into account in the understanding of the present. The combination of these different points of view gives new strength to this hypothesis of a Universe and a humanity which are not simply the results of "chance and necessity".

These new reflections update Frankl who was convinced that man, thanks to his free will, is dedicated to the realization of a future endowed with meaning. They seem to me to make credible the hypothesis of a goal to be taken into account in our vision of ourselves and of the world. But these reflections, multiple and not always concordant, are sometimes difficult to apprehend for the ordinary man that I am, and even more difficult to describe simply. Yet I became involved. I have therefore tried to write them in such a way that they are readable and interesting for the greatest number, including my grandchildren as long as their parents assist them to some extent.

I was challenged one day by this philosopher who noticed how our questions are sometimes so beautiful and the answers so often disappointing. He advanced as an explanation that the quest is beyond us while the answers that we can formulate are only up to our measure. If the reflections that follow can help those who are still in the action, the construction of a life or the education of children, and who hear or formulate the same quests, to help them forge answers a little less embarrassed, then they will not have been useless, and I would be content.

INTRODUCTION

In the second half of the 19th century, science seemed on the verge of a final accomplishment given the numerous and important discoveries that followed one another. Of course, there was still a lot to know, but physics described the Universe and how it worked so clearly that it was reasonable to assume that the essentials were close to being understood. The principle on which science was based was verified: everything that exists in nature is explained by something that exists in nature.

Newton had shown two centuries earlier how much the Universe was an immense mechanism determined by the mathematical law of gravitation. Scientific knowledge was then established as certainties. The immense world had always been there, identical, and following those laws. The question of its origin therefore no longer existed. Living beings appeared and developed on Earth by chance and natural selection.

Everything was made up of matter on which forces acted. And there was no longer any need for anything other than matter and forces to decipher the Universe. No other explanation was necessary to conceive of the world as a whole. Referring to another level of reality, to a transcendent dimension, to divinities, to a God, to a spirit, had become useless.

At the same time that science disenchanted the world, it revealed to man how mistaken he had been in thinking of occupying a special place in it. The Polish astronomer Copernicus had already, two centuries before Newton, depreciated the place of Earth by showing that it was not at the center of the world. Then Darwin, in the middle of the 19th century, showed us that man was only one animal among others, only more evolved.

The discoveries of the beginning of the 20th century further relativized the place of man in the Universe. In 1924, the American astronomer Edwin Hubble established the existence of other galaxies. Not only is the Earth not at the center of the world, not only is the Sun just an ordinary star in our galaxy, but our galaxy itself is just one galaxy among billions of others.

The emergence of life, intelligence and consciousness, attributed to chance, was no more than an accident in the long march of the Universe. And for Freud, not only was man not at the center of the world, but he was not even master of himself. Much of his actions were dictated by something of which he was unaware. Freud spoke of the triple humiliation inflicted on man by Copernicus, Darwin and himself. All three have devalued the place of human beings in their own representation of the world.

It is in this intellectual context that the Austrian psychiatrist Viktor Frankl forged a new psychological theory. It was characterized by the spiritual dimension that it attributed to man, and which was at the origin of his quest for meaning. It thus went against the current of the triumphant thesis of scientific materialism. Frankl was the first to pose the question of meaning in psychology. And in fact, it is difficult today to take an interest in psychology without noticing that the search for meaning has become an unavoidable human need.

He was obviously influenced by psychologists and philosophers who had preceded him. From their contributions, and from his tragic experience of life, he worked out a global and coherent synthesis. His was not only a psychotherapy intended for certain neuroses, characterized by the feeling of emptiness and absurdity, which he called Logotherapy. It also formed an anthropology, a philosophical reflection on man and humanity, which he called Existential Analysis.

Frankl opposed the scientific theories of his time, refusing to include man in their general determinism. And by doing so, he confronted himself with the master of his own discipline, Sigmund Freud, rallied to the scientific dogma according to which all observable facts are explicable by past causes, including in terms of the psyche.

Yet Frankl had first been a young disciple of Freud. Then he joined Alfred Adler, father of individual psychology. But he quickly distanced himself from his two masters. For him, human beings were not only in search of pleasures, as Freud asserted, and of status, as Adler thought, but they also, and above all, experienced a need to feel that their life had meaning.

This conception can challenge many of us, and in particular those who reach a stage in their life where the questioning of existence is inserted, even imposed, among the other subjects which had largely occupied them until then: to become an adult, to found a family, to raise children, to succeed professionally, to enjoy ordinary pleasures, to best overcome the difficulties, the failures, the inevitable sufferings of our condition.

This existential analysis defined a vision of man that answered questions that have suddenly become important:
- What representation of the human being, and in particular of the adult, to offer to his children?
- What are the real and deep motivations of people in their life and in its components, such as in their professional activity?
- How to exist in a materialistic, individualistic, consumerist society, which can appear so derisory to us?
- Where do these moral values come from, this conscience, which gives us sometimes this feeling of derision?
- Is it so aberrant scientifically to suppose that the physiological and psychic dimensions are not enough to completely define the human entity, and therefore that a dimension of the spirit is plausible or even necessary?
- What reflections, in the absence of answers, to bring to these questions that undeniably arise for many of us: about life, man, consciousness, spirit, the Universe: Why? Why?

Emmanuel Kant, one of the major references of Western thought, already illustrated and condensed some of them in

his own way: "Two things fill my mind with an ever-increasing wonder each time I consider them: the starry vault above me and the moral law within me". These two things are the subject of fundamental scientific research aimed at understanding and explaining the Universe on the one hand, and the human mind on the other. If science is sometimes difficult to decipher, Viktor Frankl , on the other hand, expresses ideas that are more within our reach:
- The quest for pleasure, the satisfaction of one's instincts, of one's ego, of one's thirst for recognition, are not enough to define our dimension of humanity.
- The demand for freedom cannot be separated from an obligation of responsibility.
- The aspiration to higher values, of the order of consciousness or of the spirit, is not the expression of a psychic alienation.
- The need for meaning in one's life is the emanation of the meaning of life and therefore of something that is superior to us.

This psychological theory of Frankl , widespread in Austria, in Germany, or in North America, for more than half a century, was largely ignored in France, until about ten years ago. However, everywhere it had to confront a Science that disqualified any intellectual approach calling for an explanation other than material.

In practice, materialist psychology had fairly quickly shown its limits, and this allowed the emergence, in the United States, of humanist and existential psychology, close on certain points to Viktor Frankl 's theory. Logotherapy became in fact a reference for modern psychotherapies developed in the wake of these humanistic and existential mythologies. Frankl thus finally achieved in France, a certain recognition, partly because he benefited, by association, from the significant success of these emblematic American psychologists, Rollo May, Abraham Maslow, Carl Rogers, Irving Yalom ...

And yet, Frankl 's thought deserves to be considered in itself. It carries a vision of man sufficiently developed not to remain confined to a role of inspiration of other dominant psychologies.

Beyond the pillars of his psychology, which are the search for meaning, the freedom to want, the responsibility to exist, which inspired his colleagues across the Atlantic, there is a base from which these concepts emerge. It is the transcendence of the human being, this dimension other than material, without which the other concepts collapse. How, indeed, according to Frankl , a being only made up of matter in its physical and psychic dimensions, could want to ask the question of the meaning of its existence?

And in fact, his theory is based on the integration of the spiritual dimension of man, generating his free will and his responsibility. People accepting this reference to spirituality can easily accept this theory. But those who, resistant to this dimension, or refusing to mix genres in a medical field claiming scientific rigor, can only accept this psychology with difficulty, even though it is not opposable to the great determinist and materialist principles of science.

Meanwhile, something extraordinary has been happening in the world of science over the past few decades. Meaning, consciousness, have become subjects of debate there. We thus learn that consciousness creates the sensation that we have of reality, therefore of our world.

The extraordinary advances in fundamental physics, which endeavors to understand the Universe from the largest, the cosmos, to the smallest particles, leave room for something other than matter, for lack of being able to explain the Universe exclusively by matter. This therefore opens up possibilities for new reflections in other fields, the humanities, the life sciences, psychology, philosophy, etc.

The purpose of this essay is, in a way, to do justice to Frankl who, convinced of the relevance of his intuition, opposed the dominating deterministic materialism of his time. In his reference to a specific human dimension, to an immaterial consciousness that science had then delegitimized, he finds himself in phase with the hypotheses of recognized modern scientists.

The main features of Frankl's thought are presented in the first part of this book, emphasizing how much it is fundamentally characterized by the affirmation of this spiritual dimension in man. This conviction will be at the origin, in psychology, of the first confrontation between materialism and determinism on the one hand, and spiritualism and free will on the other. These confrontations still persist.

The second part will be devoted to developments in global scientific thought. The science in which Frankl was immersed must now face, in several fields, questions without satisfactory materialist answers. The scientific principle consisting in always seeking, understanding and defining the most elementary components in order to apprehend the global, comes up against the incomprehension of what elementary matter really is, a vibration or a particle, or both at the same time.

Life sciences, cosmology, and even fundamental physics, the ultimate science of reference, are discussing the role of consciousness and the immaterial in explaining observations or results of experiments. If Frankl's intuition of a human spiritual dimension is not formally confirmed, at least it can appear today as an acceptable scientific hypothesis.

The ambition here is to clearly present the terms of the debate between exclusive materialism and dualism accepting a spiritual dimension, which has been going through scientific thought for several decades. This implies a somewhat detailed although very simplified description of the scientific theories evoked. It is indeed important to show that the debate here is not ideological but that it wants to be as rational and well-argued as it is possible to be in this kind of work.

The last part will present existential analysis reinforced by these discoveries or scientific reinterpretations, and all the more legitimate in its ambition to help us in our will to exist. It is a philosophy of life necessary for the understanding of individuals and societies today, and a source of regenerating reflections for the world of tomorrow. If the contributions of psychology help to understand man, his functioning and his motivations, existential analysis also helps to understand why man needs to understand himself and the world. It is a reflection on the human condition, on its capacity to overcome its desires, its needs, its emotions, by an aptitude to mobilize superior faculties, and on this imperative need for meaning.

This vision of the human being in search of realization of values, of exercise of his freedom in all responsibility, brought to go beyond himself, is also a vision of the man in the city. It is a political vision of the citizen, in the primary sense of the term. At a time when the place and role of man in society, no longer determined by institutions and traditions, are in question, existential analysis provides elements conducive to this redefinition.

CHAPTER I. A psychology of spirit confronted with materialism

Psychology is the last field to have claimed, in the 19th century, to constitute a science in itself, by detaching itself from philosophy. In order to achieve this recognition, it had to draw inspiration from the methods of reasoning and the development of theories that sciences had consecrated.

Science had been based on keeping the supernatural and the gods at a distance, and on explaining all natural phenomena by previous natural causes. It had developed by rejecting and denying the spiritual world. This approach, born in Greece more than 25 centuries ago, this "refusal of the miracle", this idea that everything in the world is produced by laws where the personal intervention of superior beings has no part, was implanted in all the countries which inherited the Greek culture.

However, during the centuries of domination of Western thought by the Christian religion, scholars, whether by sincere religious conviction, or by fear of the wrath of the Church, had to try to study nature while continuing to integrate the existence of an omnipotent God. Copernic, Descartes, Kepler, Pascal, explained the natural laws of a lower world, comprehensible by a human intelligence, by preserving a higher world where God organized the Universe.

But, gradually, the success of their reflections induced in the vast majority of scientists the conviction that there was no longer any need for a superior being to explain the world. The Spirit had become useless since all causes were evidently of this world.

Moreover, this astonishing ability of science to predict future events from the knowledge of past and present events was increasingly affirmed. This notion of determinism imposed itself naturally. The future of the Universe is inscribed in its past. Man, being part of the Universe, a material being in a material world, is therefore also quite naturally determined by his past.

This way of thinking, which assumes that all events can be explained by the laws of science, is the bedrock of the scientific enterprise. And it is undeniable that this will to explain facts of this world by causes of this world has allowed the extraordinary progress of science. The consequence is that any breach of this postulate becomes, in the minds of all those who aspire to scientific recognition, a fault in method, a lack of rigor, a manifestation of incompetence or dishonesty.

A) Psychology wants to assert itself as a science

When Viktor Frankl published his first articles before World War II, positivism and materialistic reductionism dominated the intellectual and moral climate of the time. Positivism is that doctrine for which only facts and scientific experience count. Materialist reductionism is this vision which reduces everything, and therefore also man, exclusively to its natural and material components. The negation of the spirit is the logical consequence.

Frankl's evolution was dominated by Freud, who had effectively extended the postulates of science to psychology. The Freudian model of psychic functioning is based on a principle of energy confrontations borrowed from the physical sciences. The individual is animated by forces in conflict, and the thoughts, the emotions, are the product of these antagonistic forces which, for the majority of them, are unconscious. The whole explanation of behavior boils down to the work of basic drives.

Psychic functioning was the product of an arbitration between an internal demand for immediate satisfaction of desires and an external pressure which refused or delayed this immediate satisfaction. This external pressure is called the reality principle. The individual, driven by his aggressive and sexual impulses, is in opposition to a world that forbids him to satisfy them. Moreover, as this psychic functioning is determined by events from the past, produced by the environment or the drives, it is essential to reconstruct the past to explain the present.

Frankl admired Freud's genius and his contributions to his field, but could not accept this rejection of the spirit. His intuition opposed it, both philosophically and psychologically, as well as what he could perceive in his patients. He could not renounce this spirituality which he considered necessary and which Freud considered useless, product of a neurosis which he called superstition or religion.

B) Frankl develops Existential Analysis

For Frankl, man is not only driven by his impulses, but also pulled by his values. He compared Freudian psychoanalysis to the vision of a certain employee of the sewers, who perceives of the city only the underground network of energy and utilities, that of affects and impulses. This employee is unaware of the city on the surface, with its universities, its churches, its temples... He therefore does not see that the impulses are primarily an energy supplying the aspiration to spiritual values.

When one proposes to deepen the thought of Frankl, one is confronted with terms, either little used in the ordinary life, or which have different meanings. It is therefore useful to try to define simply and clearly what Frankl means by words like meaning, transcendence, consciousness, spirit, existence, and their relationship to free will, and responsibility. Not that his reflection is in itself difficult to integrate, it is on the contrary clear. But it sometimes relates to concepts that we use only exceptionally in our experience of life, in our daily lives.

Let us briefly explain these terms. Their meaning will become clearer when they are integrated into the developments that follow. They are all the more essential to define as some of them have also reappeared in today's scientific terminology, such as the meaning of the universe in cosmology, consciousness in the science of life, even in the definition of reality, in fundamental physics.

- **Meaning:**

The meaning evoked here in "the meaning of life" is sometimes more easily defined by starting with the definition of nonsense, which is a feeling of uselessness, a feeling of emptiness or absurdity. It is the absence of meaning that is most spontaneously felt and therefore the easiest to define. A meaningless life is not necessarily worth living. Conversely, the meaning of life is what gives the feeling of a life worth living.

More positively, meaning includes a dimension of signification, of value. It also implies consistency with a more global context. And finally, there is a direction, a purpose, more or less clearly definable but obvious, and which can be intuitive.

Meaning is subjective. An individual perceives that his life has a meaning if he appreciates it, if he feels it as useful, animated by values, oriented towards an ideal, an absolute, or at least towards projects felt to be important.

The meaning of the universe will also be evoked, with first this component of direction, of evolution towards a finality, owing nothing to erratic chance. But it also integrates the realization of this finality into something in which we are involved, as representatives of life and of consciousness.

- **Existence:**

Man exists when he is aware of not being entirely determined, when he expresses his will to choose his life, when he assumes his responsibility to realize what he is. Existential questions are those fundamental questions about the meaning, the interest or the usefulness of one's life, which can call out to us at any time, and especially when we are confronted with crucial subjects such as death, loneliness, suffering, freedom.

Existential philosophy was forged to try to answer these questions. It thus offers reflections, paths, attitudes, likely to give meaning to one's life, and thus to help live fully. It recommends, among other things, to exercise one's responsibility to want and to act, to carry out projects that respond to the tragedy of loneliness, anxiety, despair.

Existential therapies apply to help patients suffering from neurosis or existential angst, in difficulty facing the sense of absurdity of life, loneliness in life and facing death, fear of freedom and of responsibility. They propose to lead patients, with the help of in-depth personal reflection, to want to live fully, therefore to define and implement a life that eliminates or makes bearable the anxieties generated.

We must note that this definition of existence, considered as willed and decided freely, in contrast to that of life, which is rather suffered because determined, is a convention and that this is not shared by all. Some authors reverse the terminologies for the same proofs. Victor Hugo in his long poem "Those who live are those who struggle", or Frédéric Lenoir in his works, are examples of this. They call living what we call here existing and vice versa. We will stay on the distinctions and definitions from Existential philosophy.

- **Consciousness :**

Of all these concepts, this one is undoubtedly the most difficult to explain. It has been said that we know what it is, as long as we don't have to define it. A simplified medical approach describes it as a general state of alertness and reactivity, dependent on the central nervous system and which allows the person to relate to himself and his environment.

For a philosopher, consciousness is the knowledge, intuitive or reflected, that everyone has of his existence or that of the outside world. Moreover, consciousness, beyond "this presence to oneself", of this self-consciousness, also designates the fact of being able to account for it, to speak about it.

Conscience also designates the "moral sense", this ability to distinguish right from wrong, to assess situations. Self-awareness is our ability to recognize our own life story while moral awareness is an activity that examines and criticizes; it is properly speaking an "intuitive perception of values".

Consciousness can be called passive when it is simply awake and attentive, and active when it exercises the freedom to act.

We are, however, faced with a difficulty: we have only consciousness to assess what consciousness is. And if our current knowledge still does not allow a description of the nature of consciousness, it seems impossible today to consider it as independent of the biological functions of the brain.

We will see that contemporary physics now identifies different levels of reality and invokes the notion of consciousness to define what distinguishes these realities.

- **Spirit, spirituality, noetics:**

From a philosophical point of view, spirit is defined in opposition to matter, as a substance separated from the physical world. This impalpable component is therefore different from the body. Spirituality is this ability to recognize this component.

For Frankl, spirituality, which relates to the spirit, is not belief in God, although it may include it. It covers the search for wisdom, for the meaning of life, for answers to existential questions. It is also a capacity to distance oneself from the world, including to extract oneself from material reality.

Frankl, as a scientist abstaining from any reference to God, and as a physician attached to treating everyone, including materialistic atheists, took great care in emphasizing the secular character of his psychotherapy. After initially using the German word "Geist", he found that the English and French translations by "spirit", inferred a notion of religious spirituality. He then replaced it with "noetic" from the Greek "noos", which is more neutral, since it also means "intelligence, thought". Frankl, on this subject, said he was quite close to Einstein's definition of his own spirituality. The word spirit will be used in this book to mean noos.

These different concepts are at the heart of the existential analysis developed by Viktor Frankl. He was confronted with it at a very young age, and in the first place with the question of meaning. Still a teenager, he rejected the nihilism of his time, this notion invented by the German romantics to designate this absence of any conviction, religious, aesthetic, scientific, political.

He discovered, with great interest, the work of Freud, who had considerably enriched the intuitions and hypotheses on the unconscious of some of his predecessors. Frankl was also sensitive to this therapeutic approach which contrasted with those carried out until then, by a very human behavior, where one receives a patient in a protected, safe environment, in his office. But he broke away from it fairly quickly, unable to adhere to this psychoanalysis which he perceived as too reductive.

He then turned, at the age of 19, to a former disciple of Freud, Alfred Adler. He had rejected the Freudian conceptions of intrapsychic conflicts between the authorities of the ego, the id, the superego. For Adler, man is mainly concerned with overcoming his feeling of inferiority, stemming from childhood, through mechanisms of compensation. But Frankl did not accept this vision either, still too deterministic, and which does not recognize man's freedom to will.

He was generally put off by this psychoanalysis which was inspired by plate tectonics and energy physics, and by this psychiatry which was based on chemistry. He regretted that the psychology of his time did not dare, for the most part, to integrate the spiritual dimension, which he considered to be what characterizes man. He then decided to define his own vision of Man and the resulting psychotherapy.

Frankl developed, in the 1930s, the reflections that he would finalize on his return from the concentration camps, under the names of Existential Analysis, for his vision of man, and Logotherapy in its psychotherapeutic version.

The originality of existential analysis consists in affirming that the fundamental human motivation is first of all to find meaning in one's existence, and that this need for meaning is the expression of the spiritual dimension specific to man. **In this theory, mental life is not reduced to the psyche alone; it understands the spirit. And spirit and psyche are not confused.**

The physical and the psychic are of the same nature, made up of matter, unlike the spirit. This dimension of the mind of the human being is not distinguished elsewhere as such, neither in the psychoanalysis of Freud, nor in the individual psychology of Adler, nor even in the humanistic psychology of Maslow. According to Frankl, only the human being is endowed with it, and it is for this reason that he is the only one to ask himself the question of the meaning and the value of life.

Wishing to build a work of a scientific nature, while insisting on this spiritual dimension of the human being,

Frankl based his construction on phenomenological analysis, that is to say the way things present themselves to consciousness. Edmund Husserl, one of the greatest German philosophers, is the founder of this school of thought, with the intention of making philosophy a scientific discipline. Phenomenology takes its name from the study of phenomena, of reality as it is apprehended, of experience as it is lived, of the consciousness that we have of it. It enshrines the idea that there is a separation between the "thing in itself" and the phenomenon, between the real as it is, objective, and what is perceived from it. The independent real manifests itself only by what appears. For Frankl , it is illusory to think that we can know anything about man other than what our conscience perceives. We will see that contemporary physics will return to phenomenology.

At the end of the 19th century, the science of the psyche decided to be objective, experimental, while phenomenology studied the subjective aspect of situations, and tried to reduce excessive classifications. For the latter, man cannot be reduced only to an object of observation. And in fact, phenomenology, for Frankl, is a means of describing how man understands himself, how he interprets his own existence.

Beyond phenomenology, Frankl draws inspiration from a major philosophical current, the philosophy of existence. Socrates, Pascal, Dostoïevski, are considered as its first thinkers, but it is the Danish philosopher Kierkegaard who is recognized as its founder. Martin Heidegger, in Germany and Jean-Paul Sartre, in France, will be part of this current of thought, designated thereafter under the name of existentialism. This is conceived as a reaction to the nihilism of Schopenhauer, Nietzsche, and Freud.

The main points of this philosophy of existence that humanistic psychology will integrate later are:
- Man is in the center.
- Existence is always an individual existence, but the human being is always a being in relation with others and with the world.
- Man is not determined; he must first want to realize himself, to become what he potentially is.

Frankl emphasizes that existence is not to be confused with life. Life comes under the physical and psychic dimensions, whereas existence results from the expression and deployment of spirituality, this specifically human third dimension. Man, thanks to this spiritual or noetic dimension, is no longer subject to his instincts, his impulses, his emotions, nor to the environment. He is released from the environment. It is this spirituality which allows the non-determination of existence.

Thus, a person is existential when she/he exercises her/his freedom to decide and to act. Frankl opposes human free will, the fruit of the spiritual dimension, to Freudian psychoanalysis and Adler's individual psychology which affirm that the psychic and the behaviors of each individual are exclusively determined by the history of past interactions with the environment.

Responsibility, which is the counterpart of freedom, characterizes the individual who exists. And it is he who decides before what or whom he wants to be responsible, society, humanity, conscience, divinity, ... But behind the notion of responsibility appears very quickly that of guilt. Existential guilt is the product of unrealized potentialities. It is not only the guilt of having acted, but also that of not having done so, when we have not realized our possibilities, fulfilled our potential. And how do we discover our potential? How do we know it when we are faced with decisions to be made? By the call of conscience.

However, awareness of responsibility is only the first step. The will alone allows action. Existential analysis is effective only because it frees the will of the person from all that prevents it from expressing itself. It is only fruitful if it allows him to modify his future, and to grow in maturity. The past, or more exactly the memory that we have of it, is important because it influences our current existence. But it is in the present that the future is conceived, actualized. It is therefore the present that constitutes the essential field of action.

To exist is also to seek out what has meaning, and to choose what has value. This requires a good knowledge of oneself, without being the ultimate objective. It is a transitory stage which must lead beyond oneself. This work on oneself is a self-education, in order to know how to control oneself and to be able to grow internally.

For Frankl, self-realization goes through self-distancing, this ability to gain height in relation to oneself, and through self-transcendence, this desire to surpass oneself, made possible by our spiritual dimension. Self-transcendence is the highest degree of development of a human existence. It constitutes the specifically human potential, the capacity to think beyond oneself, to act beyond the self for-itself, to exist for something or for someone other than oneself. It helps us not to fall into the trap of narcissism. And self-distancing is the first condition for moving towards this self-transcendence. It allows the human being not to be dependent on his physical and psychological dimensions.

Frankl had drawn himself once, having the size of a dwarf, and standing on the shoulders of Freud and Adler, thus wanting to underline their undeniable contribution to his own reflections. But in reality, he conceived his psychotherapeutic theory largely as a critique of their work. Beyond personality oppositions, a fundamental gap separates these currents, which evolved in parallel.

For Freud, the belief in a transcendence is at the origin of psychic suffering. For Frankl, on the contrary, it is the devaluation of transcendence, its repression or the suppression of spiritual aspiration which is at the origin of psychic suffering. Existential analysis reminds man that there is a spirit where Freud asserts that there are only drives. The discovery of the libido as a source of motivation was remarkably relevant, but the primacy and exclusivity granted to it was, for Frankl and many others, overdetermined. Admittedly, man has impulses, but it is the animal alone which is entirely constituted by its impulses.

Adler's approach, in which man ultimately seeks above all to overcome his feeling of inferiority through compensatory mechanisms, seemed just as unsatisfactory. It seemed to misunderstand too much the possibility of transcendence that exists in man, which confers on him a capacity for free decision. Existential analysis can only register in opposition to the mythologies denying the dimension of the spirit.

Frankl did not deny science and its deterministic explanations. But he refused the systematic determinism which seemed to him to proceed from a scientific ideology. One of the purposes of existential analysis is to help individuals to know themselves and to affirm themselves as persons, therefore to lead them to become aware of the determinisms which can hinder their quest for meaning, and of the possibility of overcoming them.

C) Materialist determinism and spiritualist free will

The materialist-determinist current will continue in theories such as sociologism, which explains man through social factors, and biologism, which emphasizes the determining weight of heredity in the constitution of individuals.

Behavioral psychology, or Behaviorism, will also come to claim that the "environment" entirely conditions human beings. This behaviorism wanted to make psychology a rigorous science, sticking to "objective" observable facts, and refusing any reference to consciousness, to representations of the subject, to introspection. This objective psychology attempts to identify laws where actions depend only on physical or chemical factors. For it, consciousness is not a determining factor in human behavior.

Today, neurophysiology explains the totality of the human being by the functioning of his nervous and neuronal system. All of these sciences which strive to explain man by one or a few deterministic causes illustrate what Frankl called reductionisms. These, affirming that "the human being is nothing other than the sum of his determinisms", deny to man a spiritual dimension, and the share of freedom that this dimension gives him.

And yet, psychoanalysts, trained in the Freudian tradition in Europe, then emigrating to the United States, had begun to question this notion of integral determinism. A number of renowned European psychiatrists (Binswanger, Boss, Minkowski, Kuhn, Caruso, etc.) refused the mechanistic concepts of Freudian psychoanalysis and this model of human behavior, taking an interest to one degree or another in these themes of freedom, responsibility, need for meaning.

Even among the most brilliant and closest students of Freud, such as Jung or Rank, there appeared opponents of this vision of human nature who moved away from it, in search of a specific human dimension which they sensed. Otto Rank emphasized, in the 1930s, the importance of the will. For him, this was not a secondary function as Freud and Adler considered it, but played a central role in the development of the child and in therapy.

Karen Horney insisted, in the 1940s, on the crucial role of the future in the nature of behavior. The individual is more motivated by intentions, ideals and goals than he is determined by past events. For her, too, the main task of the therapist was to free the will.

In the same period, still in the USA, psychologists, the most famous of whom was Abraham Maslow, developed in opposition to the two dominant schools, "behaviourist" and "analytical", a new thought, "humanistic psychology". They no longer accepted the rejection of what seemed to them to be major characteristics of the human person, such as values, free choice, love, creativity, self-awareness, human potential.

In 1960, a school of cognitive psychology was created at Harvard with the ambition of making a place for the spirit in the field of human sciences. It wanted to create a new psychology that considered the human being as a producer, beyond logical and abstract thoughts, "of dreams, ideas, culture; like an artist who creates, a believer who prays, a child who discovers the world…".

In fact, in psychology, the materialist/determinist current is confronted with a spiritualist thinking, itself divided on the constitution of this spirituality. Some assume that the evolution of the human brain has produced a need for concepts such as consciousness, freedom, will. These concepts induce that men behave better when they believe that their life has meaning. It is therefore logical, since effective, to adhere to it, without there being any need to go beyond. The others consider it impossible to find meaning in their life if it cannot be linked to a transcendence, of which it would only be a form of declension. They deduce that the concepts in question, consciousness and mind or spirit in particular, are not a production of the biological brain, but a component of another nature. What light do the reflections from the latest scientific advances shed on these different hypotheses?

CHAPTER II. Scientific materialism is questioned.

The scientific thought of the 18th and 19th centuries is well illustrated by the French physicist Pierre-Simon Laplace who affirmed that the hypothesis of any supernatural entity was useless in the explanation of the Universe. This position endorsed the legitimacy of an exclusively rationalist science, detached from any spiritual influence. This dogma of scientific thought could put in difficulty scholars who were suspected of being inspired by their own spiritual convictions. Georges Lemaître, an eminent Belgian astronomer and physicist of the 20th century, was the first to postulate the Big Bang theory. This idea, now recognized by all, was long refused and mocked because it emanated from a scholar who was also a catholic priest and therefore suspected of nourishing the biblical creationist theory. If a scientist was accused of creationism, of believing in the existence of an "origin", he could be excluded from the scientific community.

Only a genius like Einstein could, thanks to the unanimous recognition of the superiority of his scientific thought, refer, without rejection, to a divine intervention, behind which he did not put any personal god. The intelligence he perceived, in the organization and functioning of the Universe, amazed him.

Thus Laplace, justifying the non-existence of God by his uselessness in understanding the world, and Einstein, justifying the existence of a superior presence by the incredible intelligence of the world, represented, in a certain way, the two currents which animate, even today, the scientific community.

In fact, the great discoveries made from the Renaissance until the dawn of the 20th century had confirmed in a striking way that all reality was explained by some other reality. The "surreal" was discredited altogether.

But, during the 20th century, great thinkers, both philosophers and scientists, began to evoke the existence of other possible causes, of another level of reality, of an unknown which escapes us, necessary for explaining and understanding the world. And the scientists who have most shaken the materialistic certainties of the scientific body could only belong to the most emblematic of scientific thought, the sciences of physics and mathematics.

A) From the immensely large: Cosmology.

It took Newton, and his theory of universal gravitation, to show that everywhere in the Universe, bodies exert on each other, forces of attraction which depend on their masses and their distance. And he provided the mathematical expression to explain it. For the first time, the fusion of terrestrial mechanics, that of movement and free fall, which Galileo had largely constructed, and of celestial mechanics and gravitation, took place. Newton thus showed that the laws of physics that he had just stated applied on Earth like in Heaven.

Other physical theories followed, including that of electromagnetism, in which the British physicist James Clerk Maxwell had merged in 1865, electricity, magnetism and light. Around 1900, all these scientific laws which today are grouped under the name of classical physics, explained the nature of the world better and better and gradually answered all questions. Some scientists thought that they were on the verge of being able to explain the whole of reality, but then some disturbing questions arose.

In 1887, two American astronomers, Michelson and Morley, measured the speed of sunlight in different directions. According to Galileo and his law of speed composition, we should have found different speeds of light depending on the directions of propagation of its source, given the movement of the Earth in its orbit. Indeed this law says that the speed of the measured object is added or subtracted from that of its source. The speed of light should therefore vary according to the direction in which it is measured. Now the Michelson and Morley experiments showed that this speed was always the same in all directions, no matter the motion of the source.

If the speed, which is a ratio between a distance and a time, is invariable when the distance varies with the displacement of the measuring instrument, then the time reference must vary. Questioning the relativity of the speed of light called into question the vision of space and time accepted at the time.

Einstein adhered to their presentation, and posited that the speed of light was not relative in absolute space and time, but on the contrary, was absolute, and it was space and time that were relative. He submitted to the new vision of reality that this engendered, and reconstructed physics. He thus changed our representation of the Universe by establishing the theory of special relativity in 1905.

After light, Einstein became interested in gravitation. Newton's law that all bodies attract each other seemed mysterious to him. It is in fact exerted between two bodies without any contact between them or any material interaction, and instantaneously. Who says instantaneous says infinite speed, which seemed unacceptable to him. He advanced the revolutionary suggestion that gravity was not a real force, but a local manifestation of a curvature of spacetime; that is to say the consequence of the fact that space is not flat, but curved by the masses and energies it contains. What we continued to call the gravitational force actually results from the action of the gravitational field created by a mass.

Earth and the other planets are therefore not made to move in orbits around the Sun because of a force called gravitation, they are guided along a trajectory determined by the distorting presence of the Sun. They actually follow a straight path in a curved space. The sun's mass bends space so that, although the Earth for example follows a straight path in four-dimensional spacetime, it appears to us to be moving along an orbit in three-dimensional space. In fact, the orbits of the planets predicted by Einstein's theory are exactly the same as those predicted by Newton's theory of gravitation, with the exception of that of Mercury, the planet closest to the sun. Its orbit being more in conformity with the laws of Einstein than with those of Newton, the theory of Einstein was thus validated and space became deformable in our mind. Our intuitive perception of reality had therefore been wrong again.

The sun bends space and therefore light. In effect, the fact that space is curved means that light can no longer appear to travel in a straight line through spacetime. Thus, general relativity predicts that light must be deflected by gravitational fields. This means that light from a distant star passing close to the sun is deflected, causing the star to appear in the wrong place to an observer on Earth. This deviation has been confirmed many times over.

Moreover, his theory having endorsed an absolute speed of light, Einstein deduced that nothing can exceed this speed, that no particle can go faster than light. From the moment when we can no longer exceed a certain speed of transmission of information, we clearly perceive that two observers located at different distances from an event will not see it at the same time. The notion of present is no longer relevant. An observer can see in the present what will be future for another, and we could see as past what will correspond to the present for the other observer. It becomes impossible to agree on the concept of present.

From now on, we can no longer consider time as a great cosmic clock that punctuates the life of the Universe. For a century now, we have had to conceive of time as a local event: each object in the Universe has its own passing time. Relativity is contrary to our intuition which is forged by habit, and by our inability to perceive reality.

At the same time, the Universe turned out to be extraordinarily larger than previously thought. In the 1920s, given the limitations of observation devices, the Universe was estimated to be reduced to the Milky Way, our galaxy. But in 1924, Edwin Hubble, the American astronomer, discovered, thanks to a new telescope, that the great Andromeda Nebula, supposed to be composed of dust or gas, turns out to be formed of stars, and to constitute another galaxy. Gradually, the Universe turns out to be immensely larger than we thought. And only a fraction of that is observable, given the finite speed of light. But this part, called the cosmological horizon, contains some 2000 billion galaxies, within a radius of a few tens of billions of light-years.

This observation initially engendered the feeling that life, and therefore human beings, occupy only an infinitely negligible place in this whole. For Jacques Monod, the French Nobel Prize in Physiology, "man is lost in the indifferent immensity of the Universe".

However, astronomers find that the conditions necessary for life exist in multiple places in the Universe. Habitable planets, where liquid water can exist, are not the exception in the cosmos, as was first assumed, but the rule. Nearly half of the stars are likely to harbor a habitable planet. Within a radius of 15 light years around us, or the suburbs of the sun, more than 400 stars can support approximately 150 habitable planets. Extrapolated to the level of our galaxy, this number becomes 200 billion stars and 80 billion planets. The Universe, made up of billions of galaxies, is thus home to billions of billions of planets likely to be endowed with liquid water, and therefore to see life emerge. **These estimates show not only that the conditions for the appearance of life are not unique and improbable, but that they are on the contrary very widespread.**

As early as 1922, the great mathematical constants of the Universe had been studied, these numbers are the foundation of physical reality, independent of any unit of measurement. It was realized that the constant which regulates the behavior of the electromagnetic force, is in deep relation with the speed of light, and the charge of the electron, as well as the constant of Planck, and even that of the number Pi. To many physicists, the value of this constant cannot be due to chance; it stems from a law of nature which participates in the organization of the Universe.

And in fact, the Universe, its constitution, its evolution and its functioning, are based on a few tens of numbers fixed from the first moment of its appearance, invariable in time and space: the speed of light, the charge of the proton and the electron, the mass of the proton, of the neutron, of the electron, the mass-energy density of the Universe at the origin, the speed of expansion of the Universe at the origin, etc.

Now, knowing these laws and these constants, astrophysicists have simulated the history of the cosmos and the stages of the evolution of the Universe on computers. The results of these numerical simulations very correctly reconstruct what we know about the past and the present. The Universe is cooling and expanding; galaxies and stars form as observations have described.

Then, as an exercise, they arbitrarily modified the numbers that characterized these laws and the constants to see the effect on the simulation. Amazingly, the slightest modification of only one of these numbers makes the Universe sterile, unable to accommodate complexity and life. If the rate of expansion of the early Universe had been 0.1% faster, the current expansion of the Universe would have been 3000 times greater, preventing galaxies from forming. If this initial speed had been lower by 0.1%, the Universe would have collapsed without the creation of stars.

If only one of these numbers had been different by a tiny quantity, then the Universe would not have been constituted; it would be a primitive chaos, without stars, planets, matter, space-time, or life obviously.

Furthermore, when the Universe reached 75% of its current size, an acceleration of its expansion took place. If this acceleration had taken place a fraction of a second earlier, there would be no planets and stars. We can thus list a large number of extraordinary coincidences, and when we see these fabulous settings of the Universe which made its evolution possible with, at the end, the appearance of life, we can no longer believe that this is the result of chance. **We exist today thanks to elemental forces and initial conditions, dating back 13.8 billion years, and adjusted with extraordinary precision.**

The famous English physicist Stephen Hawking, resolutely materialistic, himself pointed out that: "the laws of physics contain many fundamental numbers, and the remarkable fact is that the value of these numbers seems to have been finely adjusted to make possible the development of life".

Einstein's theories, the revelation of the immensity and complexity of the Universe, and the surprising precision of the laws of physics, making the appearance of life inevitable, do not necessarily make it possible to question materialism categorically. But they have nevertheless led certain scientists to put forward the hypothesis that the world could result from a project, from an intelligence that escapes us. And inevitably, when we hypothesize a project, we ask the question of a creative will, not necessarily in terms of Divinity as man has supposed for millennia, but as something higher than the origin of the laws of nature, of its constants, of its destiny. These discoveries also led to the conviction that we are very far from a complete and ultimate knowledge of the Universe, far from what science thought at the end of the 19th century.

B) To the infinitely small: Quantum physics.

At the same time as the vision of the cosmos was being renewed, discoveries in particle physics were upsetting our knowledge. In 1909, the New Zealander Ernest Rutherford discovered that at the center of the atom was a nucleus which made up almost all of its mass. The volume of this nucleus is extremely small compared to the size of the orbits of the electrons which are approximately one hundred thousand times larger. If the nucleus had been one centimeter in diameter, the electrons would have been one kilometer away.

This model, inspired by the movement of the planets around the sun, is actually inconsistent with regard to the classical laws of electromagnetism. These imply that negatively charged electrons, revolving around a positively charged nucleus, should have collapsed onto it very quickly, emitting light and losing energy. Thus, the physics of that time made false predictions about what constituted the very heart of its subject, the internal structure of matter. From its questioning emerged new physics, quantum physics.

The quantum principles are today the most fundamental that we know: they govern not only the physics of atoms, but all of chemistry, a good part of biology, the physics of solids, optics, in fact the essential of empirical exact sciences. This new physics calls into question two great scientific concepts, strong objectivity which says that the laws are independent of us, and reductionism which claims that to understand the whole, one must understand the components. That is to say that when we have understood what the particles are, we understand what the atoms are, then the molecules, then all the objects.

This quantum physics emerged in 1900, when the German physicist Max Planck discovered that energy is emitted or absorbed by matter, not in a continuous and progressive way, but in small distinct quantities called quanta, thus giving its name to this new theory.

The photoelectric effect studied by Einstein a few years later, shows that on certain metals, light produces a small electric current. Strangely, this does not happen for low frequency light, regardless of the intensity of the light source, as it would if light was only a wave. Einstein realized that low-frequency photons, regardless of their number, have lower energy, insufficient to tear electrons from atoms. His work confirmed Planck's discovery of quanta.

This quantification revealed by quantum physics also applies to time and space which turn out to be granular like light. **The Universe, which was supposed to be continuous, is therefore discontinuous. And this discontinuity generates strange phenomena.**

While the atoms were conceived as mini-solar systems, the Danish physicist Bohr showed, from observations, that it was rather necessary to imagine the electrons jumping from one orbit to another without passing through an intermediate point. It was as if a planet could only take Earth's orbit or Mars' orbit, but no orbit in between.

At that point the distinction established until then between waves, a movement that propagates, and particles, a physical object occupying a precise point, was questioned. At the beginning of the 19th century, the English physicist Thomas Young had designed an experiment showing that the meeting of two light beams could produce darkness. If a beam of light passes through two narrow slits and hits a screen, one might expect to see just two bright lines in front of the slits. However, he observed several closely spaced lines, bands of light and bands of darkness. Only the physical theory of waves, known as wave theory, explains this phenomenon of superimposed wave interference, like waves on the surface of a pond, which can combine or cancel each other out. These lines, called interference fringes, are characteristic of waves and not of particles. Since the 19th century, we knew therefore that light was a wave.

But if the energy is emitted by grains, as Planck showed, light being energy, there should normally have been grains of light as well. Einstein effectively explained that the photoelectric effect exists because the electron is set in motion by the arrival of a particle of light and not a wave. Light therefore appears to consist of particles, called photons, when its energy is measured, but to be a wave in slit experiments that generate diffraction where we measure its wave nature.

The French physicist Louis de Broglie supposed, in 1924, that what was true for light, could be true for matter. The verification of this audacious hypothesis gave rise to what many consider to be the finest experiment in physics of the 20th century, also called the mystery of measurement in quantum physics. I recommend the reader to go to YouTube and look for a talk on "the most beautiful experiment in physics" (In 2002, readers of Physics World voted **Young's double-slit experiment with single electrons** as "the most beautiful experiment in physics" of all time). This is a good approach to this subject, making it possible to better understand the following thoughts.

The mystery of measurement in quantum physics

Young 's experiment was repeated, projecting light onto a screen through two horizontal slits. In this experiment, if the top slit is the only one open, the photon can reach many places on the screen. If we close this slit and open the other, the photon can also access a number of places, including points it reached by passing through the other slit. But if we open the two slits, we can see that the photon can no longer reach certain points even though it could, when only one slit was open. In a way, the two possibilities that the photon could achieve, cancel each other out.

What is even more amazing is that you get exactly the same kind of fringes if you replace the light source with a source of particles such as electrons. As with light, when you have a single slit, you get no fringe, just a uniform distribution of electrons across the screen. One might therefore think that the opening of the other slit will only increase the number of electrons hitting each point of the screen, but, in fact, because of these interferences, this number will decrease in certain places. Electrons, like light which is both wave and particle, can diffract.

When the Young's slit experiment is carried out using, no longer a flow fed by electrons, but a weak beam of individual electrons sent one by one, the pattern of interferences is found again. It's even harder to understand. If electrons are sent through the slits one at a time, each is expected to pass through one slit or the other, and so behave exactly as if the slit it passes through is unique, giving a uniform distribution on the screen. In fact, we first observe successive impacts which disperse randomly on the final screen, which seems normal. But when we accumulate the results, we find a spread with zones of dense impacts and others without any impact, an image of interference fringes. In reality, even in this case where the electrons are sent one by one, fringes appear.

For physicists, this was an extraordinary revelation. If individual particles happen to interfere with themselves, it means that the wave nature is not just a property of a large number particle beam, but a property of the particles themselves. **Every particle seems to be a particle and a wave at the same time.**

Now, even more astonishing, when we place an observation device to find out through which slit each electron passes, we then observe that they pass randomly, through one or the other slit, and as a result, there are no more interference fringes! That is to say that when we want to observe the passage of the particle, we manage to do so, but by suppressing the property of generating interference, we suppress the wave nature of the particles. There is what physicists call "wave packet reduction". This reduction consists in the transformation of a wave, extended over a large space, to the state of a localized particle.

Whereas in classical physics, the measurement of a system shows the state of the system as it is, in quantum physics, the measurement changes the state of the system. The result we observe depends on the measurement we make. A photon will appear as a wave, if we decide to carry out an experiment highlighting a wave behavior, and as a particle if we choose to observe a corpuscular behavior.

But when is the decision "made" to be a wave or a particle? The American physicist John Wheeler proposed to delay this moment as much as possible, waiting for the photon to be already inside the measuring device to decide what to do at the end, that is to say the measurement that we will finally perform. Instead of determining the passage of the photon when it crosses the slits, we wait until the light wave of the photon has passed them. The observer chooses at the last moment, either to leave the screen to obtain interference fringes, therefore the manifestation of a wave, or to use two microscopes observing each of the slits, and where a position for the photon is observed, which in this case will show its corpuscular nature.

And we see that even if we wait as long as possible, the photon manifests itself as a wave if we measure it as "being a wave" and as a particle if we measure it as "being a particle". It is thus the choice of the observer, made after the passage of the slits, which will determine in the past, through which slit the photon has traveled, through one or two slits at the same time. The decision to behave like a particle or a wave is made when the journey is almost over. This cannot be explained by classical physics. With this experience, the notions of time and space vanish. The past now depends on the future.

It was considered, at one time, that the photon was a wave when it was not observed, and a particle when it was observed. In fact, this is a simplification. The photon occupies all possible states, wave and particle, when it is not observed. All the potentialities will be actualized, and incarnated in only one of them, at the moment of the observation.

Is it the measuring device that creates the observed value? To illustrate the question differently, we can take an example, more within our reach.

If there was no one to watch and no device to record, would a rainbow exist?

It seems logical to think so, because the reflection-refraction of light in raindrops does not depend on our presence.

But physicists point out that if two people are not in exactly the same place, they are not seeing exactly the same rainbow in the same place, so the two people are seeing two different rainbows. So logically, if there is nobody, there would be no rainbow. This conclusion can be applied to automatic observation. If there is no camera to take an automatic picture, there would be no rainbow either. At the very least, we are obliged to concede that we cannot affirm that there is a rainbow if there is no observation.

This example is very close to what quantum theory describes in the measurements of objects. In both cases, it cannot be said that the observed entity existed prior to the observation. And yet, we find it difficult to accept that it is the measuring device that created the rainbow. Quantum mechanics does not tell us that it is the observer, or the recording device, that creates the observed object. **For quantum objects as for the rainbow, it cannot be said that the observed entity existed in itself, before the observation, nor that it was created by this observation.**

The greatest mysteries of quantum mechanics arise with measurement. We must give up thinking that the world is as we perceive it around us. The phenomenon of interference between particles has upset our understanding of the structure of atoms. These are the basic units of chemistry and biology, and therefore the elementary building blocks of ourselves and of everything around us. In 2004, Young's slit experiment was carried out with a molecule of 256 atoms, thus showing that this quantum phenomenon concerns large objects that we never imagined could behave like a wave.

In the experience of measurement, there is a re-materialization that transforms something like a wave that permeates all space, into a single material point, localizable in time and space. But we do not know at what particular point an electron/wave will rematerialize when it is observed or when it arrives on the screen; this re-materialization is random. To go from one point to another, a particle travels all possible paths, being present everywhere, but with a probability, a density of presence, lesser on certain paths. This means that the electrons of an atom or molecule are potentially anywhere in the entire volume of that atom or molecule.

This phenomenon constitutes the fundamental mystery of quantum physics. It has no explanation; it just describes the nature of things. The famous American physicist Richard Feynman wrote that quantum theory shows nature as absurd, from the point of view of common sense; and yet experimental observations completely confirm the predictions of quantum theory.

The Austrian physicist Erwin Schrödinger proposed a mathematical equation to describe the wave function associated with the state of a particle. This wave function gives the probability of finding the particle in space and time.

To illustrate the principle of duality, one can use a well-known metaphor, an optical illusion consisting of the superimposed images of two faces, one of a young woman, the other of an old woman. One perceives one of the two images, but never both simultaneously. Our consciousness can only perceive one of the two states, both of the superimposed image and of the wave-particle duality.

Following the upheaval of our vision of the world generated by this mysterious measure, other strange aspects of the quantum world were revealed such as decoherence, non-separability, entanglement, indetermination, and a new conception of time.

Quantum decoherence.

While the microscopic universe is governed by quantum rules, why does classical physics apply so well in our daily lives? In fact, quantum effects, like the random behavior of particles, disappear when there are interactions between quantum particles and their environment. Physicists call this phenomenon: decoherence.

Suppose an electron is enclosed in a box. Its quantum nature means that it is not present in a precise place, but that it is distributed throughout the box, in the form of a probability of presence. To locate it, you have to open the box and measure its position. Like a mist that would condense into a drop of water, the electron then appears in a specific place. This measure created an interaction with the environment that produced decoherence, resulting in a probability of diffuse presence transforming into condensation in a specific location. Decoherence is this collapse of the wave function due to the interference of the environment in the quantum system.

To observe the quantum behavior of a system, decoherence must be avoided by isolating the system as much as possible from its environment. We can then produce interferences, manifestation of the quantum wave. This is the reason why quantum effects are practically only observed in the laboratory: they are fragile and easily destroyed by decoherence.

However, even certain large molecules, containing several hundred atoms, can manifest interference effects, therefore of the wave type, in a carefully purified environment. We can therefore isolate a quantum system to a certain extent. But as soon as there is something left, gas, residual radiation, then there is interaction with the environment. Thus, the atom will lose its quantum coherence as soon as it interacts with an environment of photons, therefore as soon as it is observed.

But as macroscopic objects are never completely isolated from their environment, they have no quantum appearance for us. This mechanism of decoherence explains the classical appearance of the world for a human observer.

In reality, the system remains in a superimposed state and it is only the perception that we have of it that makes it seem reduced. **The objects of our everyday world appear to us as classical entities only because our mental and technical faculties are too limited to perceive their entire reality.** Otherwise, the world would not have the appearance of the classical world, because in reality its deep essence is the quantum state.

Non-separability: entanglement

When two quantum systems, each represented by a wave function, collide and then separate, we see that they have then lost this property of each having its own wave function. To represent them, there is only one wave function which is common to both systems. The wave function being the mathematical representation of the properties of a given quantum system, when there is a collision between two systems, after separation, it is found that neither has the well-defined values that it had initially. This is the essence of the non-separability of systems.

The experiment carried out by the French physicist Alain Aspect, in 1981, demonstrated this quantum non-separability. We observe two electrons that have already interacted. They are said to be entangled. Then we separate them. When we observe one, we create a reduction of the wave into a particle. This causes in the other, instantaneously, wherever it is, his own reduction of wave into particle, even though we make sure that there is no possible signal between the 2 particles. There is therefore a total correlation between these particles, which implies that when we make a measurement on one of the 2 particles, we operate, in fact, an instantaneous measurement on the entire system, which is the 2 entangled particles, as one entity.

This influence cannot be conveyed by matter or energy, or it could not go faster than light, and would therefore not be instantaneous. The majority of physicists believe that the 2 particles form a single object, even when they are separated by thousands of kilometers. This action, which does not decrease with distance, unlike all known forces in physics, and which propagates faster than light, is mysterious. Aspect has verified experimentally that the Universe is woven with invisible links, that things are interconnected, that matter has an intrinsic organization that no one suspected.

After interaction, the particles constitute a unique entity. Their separation in space is an illusion. This non-separability leads us to have to modify some of our intuitive representations of the Universe, related to causality or locality, and even the reality of physical space-time. Aspect shared the 2022 Nobel Price in Physics for his work.

The principle of indeterminacy

As a consequence of the reduction of waves operated by observation, Heisenberg defined a principle of indeterminacy which says that one cannot know precisely and at the same time, the position and the speed of a quantum particle. Heisenberg demonstrated that the degree of indeterminacy of the position multiplied by the degree of indeterminacy of the particle's velocity can never be smaller than a certain quantity, called Planck's constant. This limit does not depend on the system of measurement or on the particle. Heisenberg's principle of indeterminacy is an inescapable fundamental property of the world. This principle eliminates the possibility of predicting the future of the world with accuracy, since in fact one cannot measure its present state with sufficient precision.

Quantum time and classical time

In classical physics, the past existed as a well-defined sequence of events. If we know precisely the data of the present, the classical laws allowed us to completely reconstruct the past. This is consistent with our intuitive perception of the world, with its well-defined past. On the other hand, it cannot be said that a quantum particle has followed a well-defined trajectory to go from the source to the screen. We can specify its location by observing it, but between each of these observations, the particle can follow all the paths. Quantum physics therefore tells us that, even if we observe the present with precision, the past that we have not observed is like the future, indefinite, and exists only as a set of possibilities.

In the delayed choice experiment, there is a repercussion of the reduction of the wave function caused by the observation, on the past of the photon. This experiment has the consequence of calling into question our conception of the past, since the observations made on a system in the present affect its past. John Wheeler insists that the past only exists once it has been recorded in the present.

The same phenomenon applies to any quantum system. As long as a measurement has not been made, the wave-particle superposition continues to exist as the classic indeterminacy of the past. In the delayed choice experiment, the past of the photon as a quantum system is completely determined. Either way, the photon follows both paths. On the other hand, as a classical system, and particularly in its particle aspect, the state of the photon is not determined. The fact of observing or not does not modify the quantum aspect of the photon, but it modifies the classical vision that we have of this photon. When we observe, we have the classic illusion that the photon has followed only one of two paths. By the choice we make to observe or not, we have an influence on the past of the photon considered as a classical system. Wheeler calls this "observer participation". We can make choices about the classical reconstruction of the photon's past, but not about the photon's quantum past. So, our interpretation of events that appear to be purely classical may be illusory.

We can therefore distinguish two kinds of time: quantum time, which is used to evaluate the evolution of any quantum state, and classical time, the time flow of our consciousness. In quantum time, any quantum state evolves deterministically, there is no indeterminacy. In classical time, the reconstruction of the classical past of the quantum system is indeterminate.

C) Determinism and indeterminability

All the great scientific discoveries made from the Renaissance until the beginning of the 20th century had confirmed the intuition of the first Greek philosophers: all events are the product of previous events.

Classical physics, which we still use often, is deterministic and extremely precise, on a large scale. The laws which intervene there, are those of Newton for the movement, those of Maxwell for the electromagnetic field, also concerning electricity, magnetism and light, and those of the relativity of Einstein.

This determinism describes a physical system whose future state can be calculated from an initial state. It suffices to know the state of all the bodies of the Universe at a given time, their positions and their speeds, to be theoretically able to describe the state of the Universe at any time in the future. The Universe is then a vast mechanism determined by its past. A deterministic system is therefore a totally predictive system. Current science tells us that this view of the world, our view of the world, is wrong.

In classical physics, a system is at all times in a well-defined state, and therefore the physical quantities attached to it have precisely determined values. If we know that at time 0, a particle is located at the initial position A and that at time t, it will be at the final point B, then its trajectory is defined by a principle of "least action", or "energy saving", which makes it possible to determine the classic trajectory which goes from the initial point to the final point. It is thus possible to associate a trajectory with a system, which is the set of its successive positions over time.

To describe what happens in physics, on a very small scale, we use quantum mechanics. It is based on the so-called Schrödinger equation, which describes the evolution of a quantum system. Its mathematical formulation tells us that it is deterministic. As long as we stay at its level, quantum theory is precise and deterministic.

Indetermination only appears in quantum mechanics when one operates what is called a measurement. The value of a measured quantity is then determined in a probabilistic manner. An electron, a photon, a particle, does not follow a trajectory in space, but appears, here or there, at a given moment, when it collides with something else. Where and when will they appear? We cannot know for sure. All the variables fluctuate constantly, as if, on a small scale, everything was always in vibration. We do not see these fluctuations because they are too small and invisible when we observe them on the scale of the macroscopic bodies which surround us. This quantum indeterminism disappears at the macroscopic level because the probabilities given by the law of large numbers, applied to myriads of particles, allow accurate predictions.

But if the two physics, classical and quantum, tell us that everything is determined, the future remains nevertheless indeterminable. It is so at the atomic level by Heisenberg's principle of indeterminacy. It is also in the macroscopic world where the ability to accurately predict the future has been called into question by chaos theory. This theory formulates that minute inaccuracies in the description of the initial conditions of an evolving system cause profound alterations in the final result.

The famous butterfly effect describes how a beating of a butterfly's wing modifies in a tiny but real way the air pressure in its close environment which will cause a domino effect on the nearby air molecules, which can go on amplifying and cause a storm on the other side of the Earth. This makes weather forecasts unpredictable to a certain extent, but unpredictability does not lead to indeterminism, only to the indeterminable. The fact that infinite, and therefore unattainable, precision is required in the description of the initial state of systems, condemns us inevitably to the incapacity to foresee exactly their final state. Chaos is predictable for a moment, according to deterministic laws, and unpredictable in the long term.

Our Universe is deterministic but its future is indeterminable, because it is impossible for us to perfectly measure the present in all its components, to fully grasp the complexity of reality.

D) Matter and reality

Very early on, philosophers such as Plato, thought that there probably existed a world independent of us, but that our capacities of perception did not allow us to know it with certainty. However, it seems to us that our senses capture information allowing us, thanks to our way of thinking educated by science, to define a stable and coherent whole. This set, which we call physics, describes an expanding world, with a dimension of several tens of billions of light-years. This space is a real object that curves under the weight of matter. This matter is distributed in thousands of billions of galaxies, each comprising a hundred billion stars.

And yet, quantum experience shows us that the foundations of matter are not material objects. Matter does not exist in the sense that there is no grain of matter. If we could observe the atoms of an object, we would see them sometimes here, sometimes there, in perpetual agitation. The more we go towards the infinitely small in an attempt to dissect matter, the more we realize that atoms are not material objects but something quite different, that matter is ultimately a kind of illusion, that it is composed of vacuum and vibrations of very small amounts of energy that maintain its structure. The consequence is that the description of reality by familiar concepts such as that of a grain or a force is no longer appropriate.

We thus arrive at a vision of the world where the materiality of things seems to dissolve into equations. The only sufficiently stable entities that physics can regard as fundamental are numbers, functions or other mathematical beings of even more abstract appearance. A wave function or a quantum field does not have a material existence. They are "empty of matter". They only exist as a potential allowing the appearance in reality of a material form, the particle.

The objects we know, including living beings, including ourselves, are not assemblages of microscopic objects, but combinations of elementary entities that are not objects.

Furthermore, we have long conceived of reality as divided into independent subparts. Which turns out to be wrong. Non-separability tells us that the objects around us do not exist as separate beings, but that their quantum state is entangled with that of everything else in the Universe.

Now the physical world is separated into two worlds with different laws, the quantum world and the macroscopic world. Yet we know that the border between the two does not exist, and that there is no clear transition between the quantum and the macroscopic.

E) Independent reality and empirical reality

The Universe was originally made up of a magma of something called particles at very high temperatures. How did it manage to gradually structure itself during the cooling caused by the expansion of the cosmos? Where did these laws and information come from, that produced:
- In the very first milliseconds, the fusion of these particles to form protons and neutrons;
- At the first minute, the association of neutrons and protons to form helium nuclei;
- After a few hundred thousand years, the formation of hydrogen atoms;
- After a few hundred million years, the formation of the first stars by the collapse of interstellar gas and dust matter;
- After billions of years, on a planet like Earth, the combination of carbon, oxygen, nitrogen and hydrogen atoms, to give rise to living cells;
- A billion years ago, the federation of cells to form plants and animals;
- 300,000 years ago, the appearance of modern man, endowed with the consciousness necessary to conceive of these questions?

We don't know. The science that generated this knowledge and allowed these questions was based on classical physics for a long time, which had proven its ability to faithfully describe and predict macroscopic phenomena. Classical physics was said to be objective because it was supposed to be independent of the observer.

Then quantum mechanics imposed itself, over the limits and shortcomings of classical physics. Thanks to it, we see on the one hand an inaccessible reality in which the reduction of the wave packet into particles of matter has not taken place, and on the other hand an appearance of this reality where the entities seem to be in defined macroscopic states.

We are unable to describe with our usual concepts what this new physics is showing us, but all observers agree on what they are observing. And the simplest explanation for different observers to agree on what they observe is that there is something outside of them, an external world that provokes their perceptions, according to the principle known as "the common cause".

Indeed, if quantum physics proves to be remarkably predictive, it is not able to describe the concepts it uses: curved space-time, the superimposed state of a wave and a particle, the entanglement of the states of two interacting systems or non-separability. It is impossible for us to mentally represent these concepts. Only mathematical objects can express by symbolizing them, experiences that are impossible to report in everyday language. **Thus, for the first time, we must conceive of the existence of a world that is not the one that appears to us through our senses and the interpretation that our brain makes of it.**

And indeed, if our senses give us a global vision of the world in which we live, they tell us nothing about the world as it is. For this reason, the true real is defined as independent of the way we observe it. It is thus said to have a strong objectivity.

The world we think we know is called empirical real, or phenomenal, because it is composed of the phenomena we perceive. And because we share the perception of it, it is said to have a low objectivity.

This empirical reality is structured by a certain number of simple laws but which are laws only for us. Its appearance and structures are partly up to us. Einstein wrote that "Space and time are the modes by which we think and not the conditions in which we live". Space-time is a perception, a phenomenon; it is just a reality for us, it is not reality in itself.

So what physics now teaches us is that we have to consider two planes at the same time when we are interested in reality. On the one hand, we must accept the idea of a so-called independent reality because we cannot contemplate it in itself, because its existence does not depend on our perceptions and our means of observation. We can suppose that the source of the phenomena is there and seek to know it better while knowing that it is perhaps inaccessible.

On the other hand, we must note the existence of empirical reality, which is that of things, of life, of evolution, of the Universe itself, as we perceive them, but admit that it is only our apprehension of an independent real that escapes us.

Science continues to advance in the description of phenomena, from what we perceive directly or through our devices, which is the framework in which we live and which extends to stars and galaxies. But it does not say that we have progressed in the knowledge of what exists independently of us, the real in itself. Theories are useful algorithms for predicting empirical reality, but they only apply to empirical reality.

However, there are undeniably great universal laws, such as Maxwell's equations, which have remained remarkably relevant despite variations in their interpretation by different successive physical theories. We find that the phenomena obey them, and that these laws continue to remain appropriate. It is therefore possible that something of the independent real is found in some of our physical laws representing our empirical real.

This independent real is assumed to be veiled in the philosophical conception developed by the French physicist Bernard d' Espagnat according to which a true real exists, since reality resists us. Many of our theories have actually been shattered on the facts and therefore on reality. But this hard real is not the one we perceive. The thesis of the veiled reality supposes that reality in itself has structures, that our great physical laws are emanations from these structures, and that objects are in no way things in themselves. So there are at least two reals. We construct the empirical real, and we construct not only the present and the future, but even the past. But the real in itself, even if it is only veiled, is nonetheless of a completely different order from the empirical real. And if there is indeterminism, it is a trait of empirical reality, of phenomena, of predictions of observations, and not of reality in itself.

F) Realities and Consciousness

But what is this empirical reality that we claim to construct? When we see a table in a room, we deduce that there really is a table in that room. And yet, if it were not a table but an electron, the deduction would be different. Physics tells us that it is the observation itself which puts the electron in the state where we can detect it. So when we are not there to observe, the electron is not in a state detectable by us. For us, it is therefore not in the room when we are not there. The deduction is therefore valid only when it bears on objects which appear to us as macroscopic. Reality is therefore at least partly created by observation.

If we decide to observe an electron indirectly, it is a wave disseminated in space. But if we consciously choose to observe it directly, it becomes a particle located at a single point in space. Reality is constructed in this or that way depending on how we decide to observe it. **It is our consciousness that makes this decision, which means that it is consciousness that partially creates reality.**

Our traditional scientific culture would like us to avoid the notion of consciousness to describe the world. But we have to admit that we are talking about interpretations for the experiential observer, which implies selective attention to this rather than that. The observer is therefore something which has an intention, which consciously exercises a will.

It is now accepted that we cannot access, with our consciousness, the true essence of matter, the quantum field or the wave function. This essence manifests itself in the physical world through its effects: particle detection, interference effects, etc. However, we manage to represent it since mathematical entities such as the quantum field or the wave function are representations of it. So, from the moment we build a mathematical representation of it that matches that reality, then in a way, part of the essence of matter becomes conscious.

But consciousness does not create matter, as some people once thought. No longer does any physicist maintain that the reduction of the wave packet is due to a direct action of consciousness on the system. The classical aspect of the world comes from our inability to make the measurements which would show that in reality it is not classical. It is our mental structures that demand, when we consciously observe a quantum system, that there is only one answer.

But, if consciousness does not act on quantum systems, on the other hand, it generates, by definition, the empirical real, the perceived real. The real is limited to what our human abilities allow, and ultimately, it is our conscience that determines these limitations. Consciousness does not act on reality but by the limits of observations that it imposes on us, it is responsible for the form in which this reality appears to us. The physical world evolves by being subject to the rules of quantum mechanics, and consciousness, at every moment, makes an interpretation of the current state of the world. The multiplicity of our consciousnesses contributes to the emergence of empirical reality, which is therefore the product of our collective construction.

The strange paradox is that consciousness is what limits us to a very imperfect perception of the real, the empirical real, and at the same time is what allows us somehow to sense the true real in itself, beyond what we perceive.

But who is endowed with conscience? There seems to be no clear separation among the living, between non-conscious and conscious beings. It is therefore impossible to answer this question. Any assumption that assigns consciousness to a restricted category is arbitrary. I only know that I, myself, am conscious, and it seems likely to me that other men are too, and perhaps even some of the most evolved animals. It therefore seems reasonable to me to say that beings with a nervous system endowed with a certain complexity are conscious.

But if this hypothesis of a complex nervous system generating consciousness is admissible, it is more difficult to consider that it is this nervous system itself which constitutes consciousness. How to distinguish between the nervous system, the brain, and the state of consciousness?

We have seen that it is consciousness that creates empirical reality, which is not reality but only the sensation we have of it. Neurologists and physicists point out that it is indeed relevant to say that it is consciousness rather than the brain that creates this empirical reality. The brain is made of matter and is the product of observation. However, we now know that the matter we observe is only a representation of reality. If we question observable reality, we are forced to question the brain itself in its materiality. The elements of which our neurons are made of are only empirical realities, realities "for us", for our consciousness. It is then difficult to conceive that this reality engendered by consciousness constitutes consciousness. **We must then consider consciousness as partly produced by the brain, but which is something other than the brain.**

The confrontation of physicists with the intimate nature of reality leads them to think that matter comes down to vibrations, that space is malleable and that time, in a certain sense, does not exist. The world as we perceive it, only exists as a perception of our consciousness. Some recognized scientists go so far as to consider that consciousness could play a much more fundamental role in the Universe, that it would be an intrinsic and fundamental fact of it and that as such, it could never be explained.

A stunning hypothesis was formulated by John Wheeler, one of the greatest physicists of the 20th century, a student of Bohr, a colleague of Einstein, thesis director of two Nobel Prize winners in physics, Richard Feynman and Kip Thorn. He suggested that since empirical reality is created by the observer, no phenomenon is real, in the sense of empirical, before it is observed. So how to consider what happened before the appearance of consciousness? According to Wheeler, we contribute to realizing not only the here and the now, but also the distant and the distant past. The result of the delayed choice experiment, in 1984, and another in 2007, show that the consciousness of the observer is necessary to bring the Universe into existence as we perceive it. Earth before the appearance of life can only exist in an indeterminate state, and a "preconscious", empirical Universe can only exist retroactively.

Some scientists therefore believe that the ultimate stage of complexity seems to be the development of consciousness, which alone can generate an empirical Universe.

G) Mathematics and Consciousness

Since physics surprisingly brings us to consciousness, what about mathematics, the most abstract science there is? These appeared about 4000 years ago in Babylon and Egypt, and proved to be very useful in managing these first large centralized states. Counting, calculation, the foundations of arithmetic, then the bases of geometry made it possible to understand large numbers, measure lengths, surfaces, volumes.

Informed of this first knowledge, the Greeks, from the 6th century before our era, theorized the idea of demonstration, which makes it possible to establish exact and universal results from a few accepted hypotheses. Thales, Pythagoras,

Euclid developed principles still taught in our schools. At the end of the 8th century AD, Arab scholars began to study Greek mathematics. They enriched the theory of numbers and geometry, then created a new discipline, algebra, which makes the link between arithmetic and geometry.

It was in the 16th and 17th century, that European mathematicians introduced novelties, such as complex numbers. Descartes, Pascal, Fermat, Newton, Leibnitz, Huygens, Bernoulli, Euler, d'Alembert, Lagrange, and many others, considerably enriched mathematical knowledge and put it at the service of physics, with such efficiency that Galileo noted that "the book of nature was written in mathematical language".

The Universe is understandable to us because it appears to us to be governed by mathematical laws. Its behavior can be modeled using these laws. The first of the interactions to be described in mathematical language was gravitation. Thus Newton's law shows us how any object in the Universe attracts any other object with a force proportional to its mass.

Mathematicians sometimes find a curiosity in mathematics, which may seem theoretical and useless until this curiosity appears to play a fundamental role in the explanation of nature. This is how the German mathematician Georg Riemann developed, in 1854, the concept of curved space. It is a space in which the sum of the angles of a triangle is no longer equal to 180 degrees, but less or more. Just draw a triangle on a balloon to imagine it.

It is thanks to this geometry that Einstein, 60 years later, understood that physical space and the Universe are curved. Euclid's geometry is not completely adapted to these dimensions, even though on the scale of the solar system, it remains very largely relevant. It is thanks to the geometry of Riemann that Einstein was able to deploy his equations of general relativity.

$E=mc^2$, his famous formula, established an equivalence between mass and energy. It also suggested properties that were still unknown and were only physically discovered later.

The familiar concepts used until now can no longer fully describe reality; only mathematics succeed at this. Physics used the extraordinary power of mathematics to reconstruct an enormous diversity of systems from a few principles. The discovery of antiprotons, these particles which, by collision, make protons disappear, is due to experiments of the 1950s. But these facts had been predicted earlier by theorists from mathematical calculations.

Sometimes mathematical predictions run counter to the physicist's common sense. In 1915, Einstein proposed his laws of general relativity which described the functioning of the Universe, and upset the vision we had of the world. His first version of these laws described an expanding Universe. Einstein himself refused this concept of the Universe that his mathematical formulas drew. He therefore inserted in his functions a constant which only served to make his model of the Universe more stable. But observations confirmed a few years later that the Universe was actually expanding, as he had originally computed. Einstein recognized that adding this constant had been the biggest mistake of his life. His initial mathematical formulation was correct. It was the physical representation he had of it in his mind that had misled him. Even him!

The cosmos seems, with extreme precision, to be managed by mathematics. The more our understanding of the physical world and the laws of nature increases, the more we are drawn to the concepts of the world of mathematics. Eugene Wigner, in 1960, in a famous lecture, came to speak of the "unreasonable efficiency of mathematics" in the physical sciences.

Thus, all mathematicians and theoreticians of physics think of the world as a structure precisely governed by mathematical laws. Some come to conceive of the physical world as emerging from a timeless world of mathematics.

Einstein's great friend, Kurt Gödel, considered by some, the greatest logician since Aristotle, lends strong credence to the testimonies of great mathematicians that they are in contact with a world of mathematics that is not a creation of their minds. This is a Platonic conception of mathematics: "It seems that one can refute the idea that mathematics is a creation of the human mind... mathematical objects and facts exist objectively and independently of our mental actions and of our decisions". The physicist Trinh Xuan Thuan is also convinced that the extreme efficiency of mathematics in describing the physical Universe comes from the fact that the natural laws come from a Platonic world of pure entities which manifest themselves in our material world.

Advances in mathematics would therefore be, in the literal sense, discoveries, which relate to pre-existing truths, resulting from a world of pure mathematics, which the mathematician explores. It would not be a simple construction of the human mind but a world with which we would have a privileged link. This thesis of a world of mathematical truths, distinct from the physical world, and from which we must understand the physical world, seems to imply that these two worlds are as real as each other.

According to the British mathematician Roger Penrose, Nobel Prize in Physics in 2020, human beings are, from childhood, able to establish a kind of contact with the Platonic world of mathematics. When a child is shown different collections of objects, the child manages fairly quickly to identify the notion of number from these presentations. In a way, the natural numbers are already there. They exist in a Platonic world and we access this world through our ability to pay attention to things. According to Penrose, this understanding of mathematics is different from the realm of calculus. There is no boundary between understanding and consciousness in man. Our understanding is a product of consciousness.

Penrose demonstrated, using a variant of Gödel's famous theorem, that a computer cannot discover theorems of mathematics. The most rigorously logical science, that of abstract mathematics, cannot be programmed on a computer, no matter how powerful. Penrose interprets this to mean that the processes of mathematical thought, and by extension those of all conscious thought and behavior, are not implantable on a computer. The inability to be programmed into a computer, non-computability, is a feature of all consciousness. No computer can simulate consciousness. If you can put a model on a computer, it is computable. According to Penrose, certain concepts are non-computable: judgment, common sense, intuition, aesthetic sensitivity, compassion, morality…

This probable existence of a link between the human mind and the mathematical world reinforces this idea that consciousness, the expression of this mind, is not entirely produced by the brain, even if it cannot be expressed without its help.

H) Life sciences are also questioned

The theory of evolution which claims that all living organisms originated from common sources is established. The most recognized of the theories of evolution, Darwinism, is based on the hypothesis of genetic mutations triggered by chance, and a natural selection of the mutations most favorable to survival. We are thus the fruit of "chance and necessity", the title of a book by the French Nobel Prize winner Jacques Monod. The observed result, since the appearance of life, is a gradual rise in the complexity of life. All the evidence shows that if we go back to our origins, we will find a monkey, then a fish, then an invertebrate, then a bacterium.

This Darwinist theory has had such success in explaining the different forms of life, current and past, that it has become the reference. It thus consecrates the materialist explanation of the appearance of life and of man, definitively disqualifying the old religious references to creation.

As a consequence of this theory, the highest attributes of man, like consciousness or spirit, can only be material. The current dogma of neuroscience is that consciousness is a product of brain activity and according to Francis Crick, Nobel Prize in Medicine, or Jean-Pierre Changeux, the reference to the mind is now unnecessary. "Man is nothing but a bundle of neurons".

But since the second half of the 20th century, Darwinism has been confronted with certain discoveries and reflections which raise questions. Indeed, the history of most fossil species presents two characteristics that are incompatible with Darwinian gradualism, which defines evolution as progressive and constant in order to adapt to changes in the environment.

On the one hand, most species show no noticeable change throughout their lifetime on Earth. Thus, the American paleontologist Stephen Jay Gould emphasized the extreme rarity of transitional fossil forms within a same species. This is the problem of the missing links in the traces left by the paleontological layers. The earliest known fossils of a given species are very similar to the latest. On the other hand, species seem to appear quite suddenly, fully formed.

The fact that fossils show stability of species for long periods and abrupt changes for a short period contradicts a gradualist conception of evolution.

And in fact, for 60 million years, the evolution towards man seems to have taken place in stages, in a non-gradual way, starting with the most primitive primates, the prosimians. Then came the little apes, the simians, 40 million years ago. The great apes appeared 20 million years ago; archaic men, (homo habilis, erectus...), 3 million years ago, and homo sapiens, ourselves, nearly 300,000 years ago. Each species, in giving rise to the next, seems to lose its ability to evolve. The prosimians, the lemurs, haven't changed for 40 million years; the little apes for 20 million years, and the great apes for 3 million years.

Evolution, whose content and rhythm are increasingly difficult to explain, seems to depend, not on climatic changes, nor only on natural selection, but on an internal logic, directed and not random. According to the work of the French paleontologist Anne Dambricourt, the cause of our evolution does not lie outside of us, but inside of us, in a mysterious internal determinism that we cannot yet explain, but that can be clearly demonstrated.

In addition, one wonders, as our knowledge of the beginning of life progresses, about the probability that chance could create a first elementary form of life. Darwin suggested the appearance of a form of life from a primitive soup, and which arose by chance over time. But this turns out to be impossible given the complexity of the smallest form of elementary life, the cell. This one, to live, that is to say to be autonomous, to develop and to reproduce, must have an extremely dense and organized structure. Recent research shows that the simplest form of life is incredibly complex. Going from mineral matter to the smallest component of life, a protein or a gene, seems much more complicated than evolving from bacteria to humans.

Life indeed begins with the first cell, and this requires several hundred different specific biological macromolecules. Since the 1970s, scientists have given up trying to bring life out of the inert, as it now seems unfeasible to them. A succession of extremely improbable steps seems necessary for the origin of life. Francis Crick, openly materialistic atheist, writes himself that a structure such as DNA, is too complex to have appeared by chance.

It should also be noted that the transition from inert to living has only occurred once in the history of our planet, since biology tells us that there is only one ancestor common to all living beings.

Moreover, Darwinist evolution requires time, more than it has had. Kurt Gödel already claimed that mathematics would one day show that there is not enough time since the origin of the Earth to lead to man through a process of trial and error. Specialists in mathematical modeling are now showing that the levels of complexity of living beings far exceed what can be produced by Darwinian evolutionary processes. This complexity can only be, in fact, the result of an optimized algorithm where the desired goal is integrated. That is to say, it leaves no room and time for trial and error.

Many evolutionary biologists have now broken away from Darwinism by asserting that evolution is directed. It would be, in a way, predictable when viewed over a large enough time scale.

Beyond the appearance of life and evolution proper, there is the problem of the emergence of consciousness and free will, which seems to characterize the most advanced forms of life. Man, on Earth, is its or one of its representatives.

The American biologist Gerald Edelman, Nobel Prize winner, describes two levels of consciousness, a primary consciousness to which animals would have access and a higher consciousness in which language would play a determining role. Primary consciousness allows us to grasp the consequences of actions on the environment. It guides attention to perform complex tasks. It allows to evaluate and rectify errors. And so, it provides opportunities for more complex and extensive learning.

And finally, it is necessary for the appearance of a consciousness of a higher order. The individual who is endowed with it will no longer be limited to the present but will become aware of himself, of his past actions and of future possibilities. He is no longer simply able to memorize in the long term, but to be aware of this power. We all experience it when we reflect on ourselves, on our life, on how we feel.

This form of higher consciousness requires the possession of areas of the cortex specially dedicated to language. The possibility of language, thanks to the existence of these areas of the brain, is for Edelman the major characteristic that allows the emergence of this new form of consciousness, more elaborate than the primary consciousness.

Selective advantage is not enough to explain the emergence of primary consciousness. For natural selection to operate on an apparent trait, it is necessary that the chance of genetic variations has first brought out an outline of it. But how can variations in genes give rise to consciousness? The evolutionist does not know how to answer. The Darwinian scheme does not know how to explain the original emergence of primary consciousness. It can possibly justify and only after the appeared emergence, a certain development of the conscience, a consolidation, a structuring on several levels.

The French biologist Remy Chauvin saw evolution as a program seeking to achieve itself. This program could only be the realization of more and more evolved forms of consciousness. Modern man is thus probably more intelligent than its ape ancestor; the crow than the dinosaur it comes from; the dolphin from the pakicetus. It seems that the evolution of species appears as a slow rise towards intelligence and consciousness.

Consider free will, this ability to exercise real freedom in our decision-making, identified as a major marker of consciousness. For a long time, we thought that we were free in our decisions. But scientific discoveries have gradually limited the scope of our freedom. It has been discovered that our decisions, although they appear free to us, are in fact conditioned by countless factors, our DNA, the biology and chemistry of our bodies, the culture, ideology and environment in which we have lived. It was intellectually tempting to infer that we, like The Universe we live in, are entirely determined by past events. Many scientists have taken this step. Others have not.

The American neurobiologist Benjamin Libet owes his fame to his experiments in the field of consciousness and free will. He showed that before we perform a gesture, a wave of motor preparation appears in the corresponding part of our brain. Then before the act is done, either we let the process that has been developed by our unconscious take place, or we decide to stop it. We exercise our will. The decision is made about 0.5 seconds after the start of the readiness wave. The preparatory process is therefore launched before the conscious decision of the act. In fact, voluntary movement is already programmed before consciousness is aware of it.

So free will is not an illusion. At the very least, it takes the form of a right of veto over potential acts that we ourselves have not consciously initiated. We can deduce the existence of something that imposes itself on the neural processes, and which is not, to date, precisely definable. The place where this free will is exercised is called consciousness. Sensations or awareness are obviously associated with brain mechanisms.

If the inevitable evolution of life seems to lead to the emergence of consciousness from matter, this consciousness cannot be of a material nature. Indeed, we can no longer confuse consciousness and the brain. The brain is a material network of neurons, synapses and biochemicals. Consciousness is, in part, a stream of subjective experiences, such as pain and pleasure, anger and love. We don't know how to explain how consciousness emerges from the brain, or why we feel pain in some configurations of neurons and love in others.

For Libet, consciousness is a field which does not correspond to any of the known physical fields, and which no physical phenomenon or theory correctly describes. Is this consciousness unique to us?

The Universe in its immensity, the galaxies, the stars, the planets, present everywhere structures remarkably similar to those in which we live. In the same way, in the microscopic world, we observe everywhere the same atoms, the same molecules. These similarities result from the fact that the laws of physics that govern the creation and evolution of these structures are everywhere exactly the same. We can therefore logically assume that at an intermediate level, that of living organisms, where the physical conditions allow it, the same effects appear: the appearance of life, and the development of the latter towards the highest levels of complexity, like intelligence and conscience.

For the Belgian biologist Christian de Duve, Nobel Prize in Medicine, biochemical laws produce such strict constraints that chance is channeled and the appearance of life, and of consciousness, necessarily occurred several times in the Universe: "It is in the very nature of life to engender intelligence wherever and whenever the required conditions are met". **The existence of other living beings elsewhere in the Universe seems inevitable, as would the emergence of consciousness in some of them.**

I) Science and spirituality

The emergence of intelligent and conscious beings, resulting from an extraordinarily fine adjustment of the laws of nature, appears as if the Universe and its laws had been defined specifically for it to occur. The anthropic principle that the simple fact of existing selects, among all possible environments, only those that allow the emergence of life, is a statement of evidence. But it is different for those who want our existence not only to impose constraints on our environment but also on the laws of nature. It is not only our environment, the solar system, which allows the emergence of life, but also the entire Universe, which is much more difficult to explain. **Since blind chance is no longer credible, the only rational explanation for the appearance of life is indeed that it results from still unknown laws of the Universe.**

"The Universe conspires to the appearance of the living, and life conspires to the appearance of consciousness". It becomes indeed reasonable to advance the hypothesis of a project, carried by an intelligence. For Einstein, this immensity, this complexity, the intelligence of these assemblies, induce the need for an omniscient entity at the origin of the Universe, a great organizer. The American astrophysicist Georges Smoot, Nobel Prize in Physics, wrote in 1994: "The Universe seems to me the exact opposite of a universe devoid of reason. Nature is what it is, not as a result of a random sequence of meaningless events, but on the contrary because it could not be otherwise. Its evolution has been inscribed since its beginnings in a sort of cosmic DNA. There is a clear order in the evolution of the universe". Thus, the successes of physics have introduced into science the question of a creative and organizing principle, of a Great Architect, thus shattering a taboo.

At the same time as this hypothesis is established, a strange universe is revealed, the reality of which escapes us. The classic statement "everything is matter" no longer makes sense scientifically. We must abandon the idea of a world made of things. Our representations of reality have been called into question by the distortion of space, by the new definition of time, by these particles which remain in communication despite time and space. Reality goes far beyond what we can see, touch and measure. And we cannot doubt this quantum theory, which describes something fragmented and without substance, and which is the best scientific theory discovered so far. First constructed as a theory of atoms and molecules, it gradually proved to be relevant in all areas of physics, and thus asserted itself as a universal theory.

This quantum physics brings everything back, directly or indirectly, to the notion of predicting the results of experiments, which implies that it always relates to our experience, to what we see or what we feel. Man is one of the creators of this empirical world that he apprehends, a reflection of a world that exists in itself. Bohr reconstructed what Copernicus had discredited: He replaced man at the center of his own representation of the Universe.

Dualism, the idea that a spirit separate from matter can exist, has become conceivable again since quantum physics shows that a non-material dimension of reality can exist and interact with ours. Current physics render incoherent the thesis of a consciousness **produced exclusively** by matter. Many scientists today believe that consciousness, although produced by the brain, is of a different nature than the physical and chemical components of the brain. Furthermore, as we saw, there are strong indications of a possible contact between the human mind and another level of reality through the channel of mathematics. From there, it is natural to postulate that consciousness can be linked to another dimension, immaterial, of reality.

Schrodinger had underlined the paradox created by the multiplicity of consciousnesses at work. Our intersubjective, co-constructed, but unique world is developed from a plurality of consciousnesses. How to explain it if not by a hypothesis of a unity of consciousness, of which the multiplicity is only an appearance. There would be only one spirit, which shines in each one of us. Our multiple consciousnesses take part in the emergence of empirical reality, our common meeting place. Our states of consciousness are authentically correlated because there is only one mind.

By requalifying as scientific the reference to something other than matter, quantum physics has opened up new philosophical possibilities. But these, which postulate the existence of another level of reality of which ours would only be the projection, in no way imply the existence of a divine supernatural. **Nothing in quantum physics speaks for a deity.** But the belief in a creative spiritual entity that is not a personal god is spreading in the scientific community.

Einstein believed in the intelligibility of the world, and believed that this human quest for intelligibility was not the fruit of a simple effort of the will. He saw in it the manifestation of a third level of religious experience, that attained by religion when it surpassed the first two, the religion of fear, and the moral religion. This third level, not corresponding to any human concept, consisted for him, in the recognition of the sublime and marvelous character of the order which is revealed in nature as well as in the world of thought. Einstein believed in Spinoza's God who reveals himself in the ordered harmony of what exists, and not in a personal God interacting with man. Einstein's conviction could only shake the solidity of materialist thought: "All those who are seriously involved in science will one day come to understand that a spirit manifests itself in the laws of the Universe, an immensely superior spirit to that of man ". **Thus physics fiercely separated from metaphysics could henceforth be considered as a possible introduction to metaphysics!**

CHAPTER III. Existential Analysis and the Rehabilitation of the Spirit

Einstein's vision inspired that of Bernard d' Espagnat who assumes the existence of a reality towards which the human mind can and wants to tend, while being aware of the insufficiency of its capacities to access it. It also joins this representation of the existential man of Viktor Frankl, a being of expectation and research, whose nature is to tend towards something that he will never be able to reach and who, therefore, participates in metaphysics.

This evokes in a certain way Descartes who represented the world as filled, on the one hand with objects, and on the other hand with subjects endowed with consciousness, which enabled them to perceive objects. This awareness was also what set the subject free. Then came Marx and Freud who gained part of their celebrity by proclaiming "the death of the subject", by affirming that the behaviors which one believed to be free are in fact determined by social factors for Marx and by the unconscious for Freud.

Thus, this great philosophical question of the subject, free thanks to his conscience, was decided by these new sciences which explained to us how much we had the impression of making choices in complete freedom but that it was only an illusion.

It is undeniable that our decisions depend on the circumstances, on the context, on who we are. We know how to be the product of the innate and the acquired.

Nature is written in our genes, determined by the genes of our parents. And the human sciences tell us how our upbringing, our environment, our past emotions determine us, often without our being aware of it. The life sciences had also explained to us that our thoughts, our consciousness are only the product of physico-chemical processes, thereby determined by previous causes.

It should also be emphasized that the psychotherapist adept at this deterministic scientific position finds himself confronted with a serious question: without free will, where is the seat of responsibility, of autonomous will? How then to offer the patient a liberation, when one cannot change the past which entirely determines his behavior?

The problem of free will can also arise as a moral question. Assuming that men are not free to will, induces an alienating and hopeless vision of humanity because it implies that we are all playing a huge comedy of freedom. Moreover, very logically, without real freedom, one cannot invoke any responsibility. However, only a moral decision implies responsibility, therefore free will, makes the world livable. This decision of responsibility is in fact made by almost the majority of humans, if only for convenience, but while continuing for some of them, to deny the free will that goes with it.

However, if we can admit that we are very largely determined by what we are, we can see that our freedom, while not being absolute, is not zero. The notion of free will is opposed to all the determinisms of the last century, and in the first place to behaviorism and the original psychoanalysis.

A) Emergence of the philosophy and psychology of freedom and responsibility.

Materialist psychology at the start of the 20th century quickly showed its limits in the face of patients confronted with malaise, a feeling of absurdity, the emptiness of their existence, the denial of their spirituality, of their humanity. It was necessary to appeal to concepts from philosophy, to better understand these new neuroses, this malaise, these psychic pathologies which resulted from it. The therapists had to invoke notions which all require the pre-existence of a consciousness, even of a spirit, in order to manifest themselves: free will, responsibility, self-realization or actualization, the meaning of life.

However, there is still no consensus on this vision of human freedom. An important intellectual current, composed of materialistic scientists and philosophers, refuses the freedom of man, invoking the evidence of psychic and biological determinism. It is true that they are reinforced by psychologism, for which man is a simple biological product animated by unconscious motives and affects, and which today receives the support of neurosciences. This opinion is in fact becoming widespread, even dominant, in the human sciences. These currents of thought, present in psychology, philosophy and neurosciences, want to eliminate the concepts of conscience, intentionality, spirit, in a vision where the brain would be the cause of all psychism.

There is, however, a paradox in most for these supporters of determinism who most often recognize the evidence of responsibility in facts, for individuals to whom they deny freedom in principle. Very few forgive the crimes of which they may be victims, which they should nevertheless do in the name of the determinism of the perpetrators. They, like all of us, make moral judgments about others, whereas logically they should consider them as conditioned individuals. They most often display their morals, their ethics, their virtues, as the results of personal merit. However, it should then only be the fruit of genetic and environmental history and the physical, psychic, cerebral dispositions which result from it.

Basically, to be responsible, you have to be free. Why deny this freedom and accept the notion of responsibility, which is conditioned by freedom? It seems difficult for many to renounce the ideology of deterministic materialism, that is, to attribute freedom to a purely material being, totally subject to the constraints of matter and conditioning. Because to introduce freedom, it would be necessary to introduce a non-material, therefore spiritual authority, whereas those same people have posed a priori that this authority could not exist.

Faced with this position, the philosopher Jean-Paul Sartre and the psychiatrist Viktor Frankl, both rich in their existential convictions, and moreover in opposition on other subjects, shared a common point: the affirmation of the radical character of freedom, and the critique of psychologism, which constantly invokes the weight of determinism. Sartre's vision of freedom is indeed categorical: **"Not only are we free, but we are condemned to be free".** And because we are free, we are responsible for our lives, not only for our actions, but also for our inactions.

Heidegger and Sartre had long studied the sense of responsibility for human beings. Responsibility is inextricably linked to freedom. The concept of responsibility only makes sense if the subject is free to constitute the world in one way among many others. We are responsible for what we do and what we choose to ignore. Sartre, moreover, does not situate himself on the level of morality. He doesn't say we should do otherwise, just that what we do, or don't do, is our responsibility.

However, it is tempting to escape this responsibility. According to Heidegger and Sartre, we seek to construct denial: "we constitute the world so that it appears to us independent of our constitutive act". When we allow ourselves to be taken in by these subterfuges which allow us to flee our freedom, we live in bad faith for Sartre, or in an inauthentic way for Heidegger.

Sartre's ambition was to free us from bad faith, by helping us to assume our responsibility. It was also Frankl's project to inspire us to fully commit to fulfilling our lives. Existential reflections agree on the duty to get involved in life, through activities that seem right and good.

With Sartre and Frankl, the existential current was far from seeing the death of the subject. The American psychologist Rollo May, on the contrary, defined existentialism as the endeavor to understand man by going beyond the split between subject and object, which, according to him, had poisoned Western thought and science since the Renaissance. This existential theory called into question the traditional Cartesian vision by considering that the person is not a subject who perceives an external reality, but that he is a consciousness endowed with freedom and who participates in the construction of reality.

The new psychotherapies now agree, for the most part, on the importance of personal responsibility. And it is not by chance; the therapies reflect the pathologies they must treat. Frankl shared this opinion that each founder of a psychotherapeutic theory describes, in fact, his own neurosis, his own history, and the society in which he lives. He thought that Freud, Adler and himself were illustrations of this.

Vienna at the beginning of the 20th century, where Freudian psychology developed, was steeped in Victorian culture. Impulses, especially sexual ones, had to be suppressed. Social rules were very restrictive. Freud understood that such strong repression of natural instincts could only be harmful to the psyche. The energy of the libido, that instinctive search for pleasure, which was not allowed to express itself freely was dissipated by indirect modes of expression. And these constituted in fact all the symptoms of neurosis at the time.

Today, many instinctual desires can be expressed freely, and sexual permissiveness begins early in adolescence. Several generations have now experienced this regime of permissiveness. Traditional or structural limits of all kinds have been lifted. As a result, psychic pathologies have been transformed. Today's patient struggles more with the feeling of freedom than with the repression of impulses. This patient, who is no longer pushed from within or prompted from without by what he should do, is now grappling with the difficult problem of choice, with what he really wants to do.

Our deep nature is still the same, but we are now faced with this fundamental issue of freedom/responsibility that social, religious and psychological institutions had long concealed from us. We are not prepared to face this challenge, and seek how to face our anguish. Individually as socially, we try to escape our freedom, avoiding awareness of our responsibility. We then put in place strategies of delegation or denial of this responsibility, avoidance of autonomous behavior, refusal of decision-making, constituting a form of pathology.

Frankl transposed the contributions of the philosophy of existence into medicine. Existential analysis considers responsibility as the very essence of human existence. And we now know that we are responsible for who we are, much more than we thought. We cannot suppress our primal instincts, hunger, thirst, sleep, sexual desire, fear... but we can control them. Our feelings, emotions, unconscious or conscious, are the product of our past, but we can analyze them and largely modulate them.

Our reflections, opinions, reasoning are by definition the result of the exercise of our will. We don't create our life once and for all, we constantly create ourselves. We exercise our responsibility continuously. Therefore, assuming our responsibility for the past and the present implies the responsibility for our future. "I am the only one who can change the world I have created". We then perceive how much assuming our responsibility is the expression of a desire to be free. But this free will cannot be the product of the brain, therefore of the material world, which is completely deterministic. It can only be generated by something non-material, the spirit, or the consciousness.

American psychologist Abraham Maslow has had an immense influence on modern psychology. He is considered the precursor of humanistic psychology, which intersects with existential psychology in many points. Self-actualization plays a central role in Maslow's conception. This self-actualization translates into the fulfillment of cognitive needs, such as knowledge, wisdom, congruence, and aesthetic needs, such as integration, beauty, creativity, harmony. For Maslow, the human being is pushed towards positive values, serenity, courage, love, altruism. Accomplished, or actualized individuals devote themselves to goals that transcend the self: global issues such as the fight against poverty or bigotry, or for ecology, or other more specific ones such as the growth of loved ones.

Frankl 's views on human motivation are close to Maslow's humanistic psychology, but were developed in a different context. Existential analysis was created by a European psychiatrist and holocaust survivor. Yalom underlines how much existential philosophy is the fruit of European reflections, confronted with nihilisms, tragedies, conflicts, sufferings. Humanist philosophy is, for its part, largely a North American production, certainly in reaction to the determinisms and limits of psychoanalysis and behavioral theories, but of a more positive, open nature. There is a tragic optimism versus a blue sky optimism.

B) An Existential Analysis enriched with contemporary reflections

Frankl's reflections can and must be enlightened by the work that has been carried out after him, both by his disciples and by humanistic or existential psychologists, or by neurosciences.

In recent decades, research has focused on our need for meaning, and confirmed it. They show how we try to spontaneously explain all our feelings in a logical diagram. We integrate our behaviors, our sensations, into a frame of reference that seems understandable to us. When we don't succeed, we are annoyed, unhappy. This irritation remains until we manage to integrate them into a larger understanding.

The American neuroscientist Michael Gazzaniga demonstrated experimentally the imperative need of man to give meaning to his actions. This experiment was done on epileptic patients who previously had the two hemispheres of the brain separated. Doctors had cut the corpus callosum which connects them, in order to suppress their most violent crises. As the language area is in the left brain, the patient can only explain through language what his left brain perceives. When asked to express what his right brain sees through his left eye, he knows how to designate it in photos, but not express it verbally. The oral response is therefore sometimes inappropriate because it is irrelevant to what he is pointing to in the photo. When asked the why of this answer, by pointing out its oddity, the patient invents at all costs a sensible reason for his answer. **This experience confirms that the question of**

meaning is so important for man that when he does not know the meaning of one of his own acts, he will immediately invent one and believe in it.

It should be noted that in these patients with operated brains, there is no disturbance of consciousness. The integrity of consciousness in a brain whose hemispheres are separated is an additional argument in favor of a consciousness not assimilable to the brain.

Faced with existential situations, we react in the same way. If it seems to us that we live in an indifferent or absurd world, then we generally experience a feeling of unease, of dissatisfaction, which we do not always manage to identify clearly. We then look for explanations that provide coherence and meaning to our existence. Obviously, we need meaning and if we cannot find it, we feel, beyond annoyance and dissatisfaction, vulnerability. Seeing meaning relieves us of anxiety, gives us a sense of mastery, and comfort.

But if the individual thinks he resides in an insane cosmos, an absurd world, without meaning or purpose, it is more difficult for him to nourish his life with meaning, since he has to create it from scratch, from nothing . **We are looking for a purpose, but feeling this purpose does not satisfy us if we think that we are the creators of it.**

The Israeli professor Yuval Noah Harari describes homo sapiens as an individual who likes stories, and particularly those in which he can insert himself, whether it is his own or that of the Universe. Seeking the meaning of life means wanting to fit into a story that explains to us what our reality is and what is our role in the cosmic story. This role makes us part of something bigger than ourselves, giving meaning to our experiences and decisions. Thus we seek meaning by inserting ourselves into a story about the Universe.

And this story is not only a reference to our history. If we have to find coherence in the past, we do not live only in the past. It is true that some psychology, and not the least, advocates that to explain the present, it is necessary to find its origins in the past, that the sources and causes of behavior must be found in the previous circumstances in the life of the person. The existential analysis of Frankl claims on the contrary that the past does not explain everything, that we think, feel, also act to achieve a purpose, a project, a freely conceived future.

In each of us, at the conscious and unconscious levels, there is a sense of finality, of the objectives towards which we tend, an awareness of a destiny, of inevitable death, which leads us to draw intentions and a behavior for the future. Thus, we must not resolve to live a future already fully inscribed in the past, but be aware of being able to implement what we aspire to, even if the entirety of its realization is not always acquired on arrival.

Existential analysis emphasizes that it is the dimension of the spirit with which each human being is endowed that allows us what Jung and Maslow have named elsewhere, the actualization or realization or fulfillment of oneself. And it turns out that today's science no longer categorically rejects this hypothesis of a spirit, no longer considered as a gift granted by a personal god to men, but as an emergence of the Universe and of life. A man endowed with a spirit constitutes the most advanced degree, in the world which is accessible to us, of a manifestation of life whose program derives from laws coming from elsewhere. **The imperious need for meaning of a man endowed with conscience, freedom and responsibility, tending towards values and transcendence, no longer results from the will of a personal God but from universal laws which govern us all.**

And indeed, many scientists today consider the transcendent dimension of the mind as a necessity. Man cannot be reduced to his biological dimension. He is also a subject in search of meaning and inner freedom, endowed with a capacity to think about himself, to determine himself, and to evaluate himself. This ability, referred to as consciousness, is a manifestation of the mind that appears to be housed in, but separate from, the brain. Since its characteristics cannot be applied to matter, consciousness cannot be considered as produced by matter-energy, or the brain. **The brain is material and conditioned, while the spirit is immaterial and free.** This contradiction has never been resolved until now and, for this reason, the status of consciousness remains undetermined.

The notion of consciousness has several degrees. The first is defined in opposition to what is unconscious. But the unconscious represents on the one hand, processes that are implemented in us, without our deciding it and identifying it explicitly, and on the other hand, this place where emotional memories reside, to which we cannot access because stored implicitly. Much of what we do is unconscious. That doesn't mean we do it for no reason. During our lifetime, through our experiences, we develop a personality that manages most of our daily interactions with the world. It is we who have developed this unconscious. We need to get to know it as best we can. The famous Austrian psychoanalyst Otto Rank, a former disciple of Freud before being expelled from the movement, wrote: "it is astonishing to see how many things the patient knows and how relatively few things he is unaware of, if he is not provided this convenient excuse for refusing responsibility".

Beyond the unveiling of what is unconscious, and the exercise of free will, making it possible to master this unconscious, consciousness manifests a need for meaning, which seems specifically human. But there again, which consciousness is it? Indeed, the work of the neurologist Libet tells us that primary consciousness does not allow free decisions to be made but just to cancel decisions already made by the brain. It is only suspensive. It must be deduced that it is only by being connected to our higher consciousness that we can think freely and can consciously make real decisions. The rest of the time, for lack of enough work on ourselves, we are only able to say no.

This superior consciousness is at the origin of this need for meaning because it is the expression of the dimension of the spirit when the psyche is the place of practical intelligence. It is also the place of moral conscience, of the judgment of values, when the psyche is in fact the instance of requirements and taboos. We perceive then how much consciousness encompasses, beyond the conscious.

Jung considered that psychotherapy cannot have the sole purpose of treating pathologies, but that it must increase the psychic and spiritual level of the individual. He thought that after a first part of our life devoted to production and reproduction, we should devote the second part to realizing ourselves. To fully accomplish our existence, before dying, we must raise our level of consciousness. Death is therefore not an end, it is a goal we must reach as full as possible.

The idea that each human being has a unique potential to realize is very old. Aristotle already spoke of it. Almost all contemporary humanist or existential theorists or therapists refer to it. The subject who does not live fully can experience a deep and powerful feeling that can be described as existential guilt.

Existential emptiness, or its attenuated form, existential frustration, the product of an insufficiently endowed existence, is not a disease. Frustrations and existential crises are the expression of sincerity, lucidity, and can be beneficial. They are the expression of a life in need of coherence.

Many of today's behaviors stem from manifestations of the existential void, already cited in his time by Frankl:
- Complaints about emptiness, absurdity;
- Lack of projects, initiatives; motivational disturbances;
- Aggressivity of angry people, always on the verge of exploding;
- Boredom, fatigue, indifference, lack of interest, apathy;
- Addiction to alcohol, drugs, or depression.

- Constant search for sexual pleasure and desire for power;
- Escape from oneself through professional hyperactivity or leisure so as not to confront the emptiness that inhabits us.
- Tiredness of being oneself...

We know how to be the object of deeply rooted impulses. The role of civilization is, at a minimum, to promote positive impulses, such as mutual aid or respect for others, and to curb negative impulses, such as envy or aggressiveness. But it must do more, and bring human beings to master their instincts, their impulses, their emotions, by becoming aware of the superiority of a specific dimension, that of the mind. Frankl adheres so much to this vision that he places existential analysis at the service of the elevation of the consciousness of Man and humanity.

C) Existential Analysis elevates the individual into a person

It highlights the gap that can exist between an individual who is content to live and a person who exercises his will to exist. The existence of man is situated in the opposition between his nature and his spirit, which Frankl calls psycho-noetic antagonism, between the psyche which is matter and the spirit which is not. It is the mind that unties man from the physiological and allows him to respond to the need for meaning. Our instincts, our primary emotions, have an overriding force that can enslave us. We must use our conscience to master them, to channel them in order to continue to feel this energy without letting it express itself wildly. To follow one's instincts without control is to abdicate one's responsibility and one's freedom. **The spirit leads to openness to the world, when the psyche encourages the preservation of the interests of the ego.**

It is this dimension of the spirit which makes the human being able to emancipate himself from the determinisms which weigh on him. Jung had described the process of individuation, of self-realization, as a spiritual journey. Individuation is not individualism, which induces an attitude of affirmation and self-preference. On the contrary, it is about including the other and the Universe. It is a process of psychic growth which imposes itself quite naturally on the individual. It arises from a deep need for meaning or a change of meaning, following suffering. This very often corresponds to the existential crisis of the middle of life.

In a society that encourages man, above all, to be happy, Frankl emphasizes that it is difficult for man to be happy in himself. What is attainable is having a reason to be happy. As soon as we have a reason to be happy, happiness will most likely set in. And this reason is neither the feeling of power, nor of pleasure, but the feeling of a meaning. Frankl distinguishes the impulses, sexual or aggressive for example, which push an individual from the inside, even from below, versus the notion of sense which pulls the individual from the outside, from above. He opposes to the psychology of the depths, this psychology of the heights for which the man, beyond his impulses, aspires to the search for meaning of life, and to insert there the meaning of his own.

Frankl thus differentiates drives and aspiration. At the heart of what we are, in what differentiates us from the animal, we are not driven by instincts, but on the contrary, drawn by a goal. This aspiration implies not only that we are directed towards something outside of us, but also that we are free to accept or refuse the goal that calls us. Aspiration is oriented towards the future: we are pulled by what must happen, more than pushed by the forces of the past and the present. In an existential dynamic, everyone experiences tension between their current situation and their goal.

Frankl wrote in 1983: "the contribution of existential analysis consists in considering man fundamentally for a being who is in constant search of meaning. However, this search for meaning is proving increasingly futile in today's societal conditions… No one is spared from the confrontation with inevitable suffering, with inexorable fault and finally with inevitable death. Despite this or perhaps from this, it is a question of drawing something meaningful from it, and transforming suffering into accomplishment, fault into change, death into incentive for responsible action…".

Existential analysis is alone, in the psychological domain, in presenting existence as an expression of the spirit, and in postulating that "the human being" is first of all "existing". Man exists by devoting himself to projects that transcend him, whether by committing himself against poverty and sectarianism, or for ecology, or by helping the growth of those close to him. It exists in the action of doing or creating, in what gives meaning by experiencing something, by loving someone or by acting for others. It also exists in attitudes which, in the face of painful or desperate situations, are exemplary of dignity. He needs to feel called to accomplish something, an act of altruism or devotion to a cause, artistic or scientific creation, self-actualization by fulfilling an innate potential, self-transcendence, in turning to something or someone outside or above oneself...

These activities, which consist in making the world more livable, in putting themselves at the service of others, allow many people to find meaning in their lives, and thus even in their death. The prospect of death seems less disturbing when the individual has the feeling of having lived well.

It is introspection that allows us to become the master of our life, to learn to accept others, to develop empathy, to live in solidarity with others. The person with whom we must communicate the most is ourselves. We have to perceive that, knowing ourselves, we expand the conscious: those previously unconscious things we help to become conscious. The process of existential analysis is to illuminate the unconscious accessibility to consciousness, so that it can be managed rationally, with maturity. The less mature we are, the more we are regulated by unconscious things. An immature being is essentially guided by things it does not understand.

If introspection is essential to any understanding of oneself, it is not a question of wallowing in this hyper-reflection on oneself which extinguishes any possibility of action. It is a question of identifying its faults, its inadequacies, selfishness, cowardice, indifference, enslavement to material goods, etc. We must bring these obscure aspects of our personality back to consciousness, in order to be able to correct them. It is this discernment that makes it possible to identify the real motivations, those that come from affectivity or desire, to keep the good ones and reject the bad ones. This exercise of discernment is an essential step in the process of individuation.

Psychoanalyst Cynthia Fleury identifies the loss of discernment as the first symptom of narcissistic pathologies and psychotic disorders. But discerning requires time, lucidity, courage. Indeed, this requirement is difficult to establish and maintain in a world flooded with information, too often false, with simplistic, provocative, vulgar comments, carried along by social networks. This discernment is what existential analysis calls self-distancing and self-transcendence. There needs to be a considerable deployment of moral force, of energy in appealing to this moral conscience.

Where does this moral conscience come from, which Kant compared to the "conscience of a court of justice internal to man". According to the American philosopher Richard Boyd, a professor at Harvard, moral values cannot only be creations of human brains but must also emanate from something else, present in the Universe, pre-existing to human life. The idea of a principle of pleasure which would be at the origin of our morality, and with it, of our conduct in society, can reasonably be abandoned.

For Frankl, "Awareness or consciousness, is the intuitive ability to detect a unique and particular meaning, which is hidden in every situation. Consciousness is a sense. But conscience is not infallible and one can be mistaken in the interpretation of the meaning, in the identification of the moral value to be implemented. And because listening to the voice of one's conscience entails the risk of error, an authentic existence, endowed with meaning, implies knowing how to accept making mistakes, with tolerance and humility.

Frankl held self-transcendence, the ability to go by oneself beyond oneself, for the highest degree of development in a human existence. No school of psychology thought, before Frankl, that what essentially counts for man could be outside of himself. In-depth psychology, behavioral therapy, even humanistic psychology, make of man an egocentric representation and do not account for a being who is essentially focused on something beyond his self. Existence is only possible through transcendence. Without transcendence, there is no freedom. It is, as we have seen, an opposite perspective to that of Sartre, for whom human freedom does not exist, if there is reference to the spirit.

D) Existential analysis and humanization of society

Beyond the person, what about society? The Canadian psychologist Steven Pinker set out to show how civilization had, throughout its development, globally reduced suffering and increased consciousness. Now, if it continues to reduce physiological suffering, will it be the same for psychic suffering and is the increase in consciousness still a project of civilization? The rapidly growing consumption of narcotics, anti-depressants and other similar products expresses a development of psychic malaise in our society. Social fractures, manifestations of violence, problems of incivility, insecurity, the aggravation of which is less and less disputed, and described here and there as the savagery of society, do not bear witness to an increase in the level of our collective consciousness.

Access to happiness for the greatest number seems in fact to be on the decline, even though a kind of tyranny of happiness has imposed itself in recent decades. Life must be absolutely happy, full of pleasures and satisfactions. This injunction, social and moral, is everywhere, but we can see the perverse effects of this discourse. Putting happiness at the center of one's life seems to reinforce narcissism. And this narcissistic quest for happiness stems from an individualistic vision of society. When the individual feels that his interest is different and more important than that of the group, the frenzied individualism that characterizes our time flourishes. We no longer accept the insufficiency of happiness which then causes dissatisfaction, resentment and anger, and we blame it on others.

This conviction of personal irresponsibility by transfer to the responsibility of others is one of the expressions of this great sentimental movement which currently structures society, according to the philosopher Cynthia Fleury. "Individuals childishly alternate aggressiveness and denigration. Their posture as victims resembles those well-known psychic pathologies of patients ingeniously producing the absence of a solution. Everything that is proposed has already been tried and proved ineffective; everything that has not been tried is devalued. Their arrogance is immense, without doubt the only defensive bulwark against the definitive invasion of low self-esteem".

It shows how certain political ideologies create and maintain alienating resentment. The authors and carriers of these ideologies live and enjoy it. They induce, by inconsequence or by calculation, a false and deterministic vision of the world which condemns the victims to remain so. The French philosopher Deleuze already showed how the man of resentment goes from an inability to admire to an inability to respect anything.

Frustration and resentment grow on the belief of the "right to". We had already noted that democracy inherently creates resentment because the notion of equality is a fundamental issue. But Cynthia Fleury points out that men can behave differently in the face of their resentment. They can feel bitter about it, without making it a victim status. **Societies, like individuals, decide whether or not to let themselves be dominated by their impulses and their delusional victimhood.**

Today, ideologies, individualisms, insufficient education, do create and maintain resentment, a cause of alienation for the individual and incivilization for society. The appeal to reason seems ineffective, as this resentment appeals to the instincts to exist through victimization and aggressiveness, and the bitter pleasure of irresponsibility. Adler already underlined that one of the effects of modernity is that the fact of hiding one's wounds, therefore one's vulnerability which had prevailed until then, has turned into the exhibition of one's wounds, even if it means inventing or exaggerating them, as an advantageous factor, for manipulation or victimization.

Frankl underlined how the individual in ill-being revolves around himself, flees responsibility, delegates decisions to others, then transfers the dissatisfaction of his suffered life to the responsibility of others. Resentment and victimization, associated with an aggressive demand for freedom, are close to a form of neurosis, which Frankl characterized as noogenic, because linked to an unsatisfied need for meaning. The difficulty of giving meaning to existence constitutes the fundamental human dissatisfaction. This neurosis characterizes our current society.

The widespread phenomena of depression, hyperactivity, aggression and addiction are the product of the existential void. To free himself from this noogenic neurosis and remain in good psychic health, man needs a specific tension towards a meaning. The problem with modern medicine is that it addresses one of the two dimensions, physical or psychic, whereas in many cases it would be necessary to address the whole person, including his third dimension, that of the spirit. Existential analysis makes it possible to limit the excessive medicalization of the problems of existence. The care of a drug addict, an alcoholic or a depressed person requires treating the three dimensions, combining the resources of medication and those of mental motivation.

We have seen that psychoanalysis had revealed to us the will to pleasure, just as individual psychology had familiarized us with the will to show off. Then Frankl represented to us the will to meaning as even more important for man. In addition, he specifies that these human motivations most often follow a sequence: the pleasure principle constitutes the guiding principle of the child; the will to power of the adolescent and young adult; finally, the desire for meaning, that of the mature adult.

We perceive how many of the evils of our society come from the insufficient maturity of adults still in search of pleasure and power. How can we promote and accentuate this access to maturity? How to mobilize the greatest number of energies to put them at the service of moral and noetic values? Education, in the broadest sense of the word, is the key to progress. Any long-term education project must emphasize ethical and civic behaviors, encourage traditional moral values, honesty, benevolence, empathy, compassion, gratitude, modesty.

The pedagogy inspired by existential analysis encompasses this project by insisting more on the way to achieve it, by the control of the drives, the empowerment, the attention, the role of the conscience and the spirit. It must be apprehended by both teachers and parents. We all need to understand how children are pleasure beings, and teenagers status beings. They will only become truly adults by mastering their impulses and emotions, by being drawn by a dimension of the spirit which will assure them of freedom and responsibility. They will access happiness by existing, and not by the satisfaction of impulses.

Pedagogical methods and behaviors should be thought out in terms of this objective. Thus, it would no doubt be relevant to appeal more, in institutions dedicated to education, to the principle of responsibility rather than to the principles of pleasure or recognition as a motivational tool. More than ever, education must be an education in responsibility. And being responsible means knowing how to resist the incessant stimuli that invade us and to make choices among them.

Teaching should be devoted not only to imparting knowledge, but also to sharpening self-awareness, attention, concentration, and refining moral conscience. Someone who doesn't know values, doesn't know his worth. A person who belittles everything, who does not recognize values in others, is most likely someone who has not found his own value, and lacks self-esteem. Education must promote a values orientation as well as the fulfillment of several values, in order to avoid the void resulting from the disappearance of a single value. The goal is not to excel in each of its values, but to experience joy in its exercise.

Moreover, education is so crucial that it cannot be delegated exclusively to professional educators. All adults have a responsibility in this area, and in the first place, parents.

The current main problem of parents is that they want to see their children as happy as possible. What parent does not experience this feeling spontaneously? But giving children the ability to find meaning in their lives is not trying to make them happy in order to be happy. It is to bring them to the ability to build fulfilling projects that are part of the future, and not through a life of pleasures in the present. Wanting at all costs the happiness of one's children in the present, without cloud, without annoyance, without frustration, will not help them to find, in the future, a meaning in their life, and will therefore affect their ability to access the happiness that would follow.

Moreover, educating one's children in the quest for meaning gives meaning to the very lives of parents.

Every adult is called upon to fulfill this primordial function of awakening consciousness. It is a question of promoting in the child the awareness of himself and of his relationships with others, in order to increase his knowledge, to lead him to help others and not to increase the misery of the world. This is what ancient wisdom already defined as the ambition of an education.

Beyond education, the economic and political models in which we live are obviously affected by a new vision of man who takes due account of the need for meaning, morality, spirit, in the face of traditional springs that are too widely used, the satisfaction of instincts, emotions and interests. It is not a question of renouncing all the components of the current models, some of which have proven and are still proving their worth. Nuisances, both for men and for nature, occur in the lack of moral inspiration that presides over their implementation, most often based on the satisfaction of material needs and recognition, giving rise to individualistic, egocentric behavior, and manifestations of aggressiveness when they are not considered sufficient.

Existential analysis can be a precious reference in the elaboration of these new models, which will be applied with all the more assent when the education mentioned has raised everyone's consciousness.

E) Re-enchanting Man and the world

The quest for meaning is the foundation of Existential Analysis. But in what world does this quest for meaning take place?
- The grandiose, fascinating or disturbing, welcoming or indifferent Universe: the real in itself that we only sense, knowing that we may never really know it?
- The empirical world, of which we have a common perception, and of which we know that it is only a phenomenal representation?
- Or the subjective world of our personality, which we can modify by ourselves?

Three levels of understanding and action ensue for this exercise of responsibility in the quest for meaning:
- The Universe which brings us back to nothing if it seems indifferent to us, but which increases us if we feel associated with its project.
- The empirical world that we perceive and represent, that we can partly modify ourselves by the direct or indirect influence that we exert on it.
- Our inner world, composed of organic matter and of spirit, driven by instincts, emotions, reflection, consciousness, and on which we can largely act to build or find meaning.

The "transcendental" project of existential analysis is not to create the world. But if Man is not the creator of the world, he is not a passive observer either. In a way, he is himself a subjective self-construction that constitutes a world within the independent world.

Indeed, if the idea that the Universe, life, consciousness, resulted from the only forces of chance and selection, may have seemed convincing for a moment, it now seems very unlikely. The contrary thesis which wants that our Universe carries a direction, even a goal is much more plausible.

Hypotheses from some recognized scientists seriously assume that the purpose of the Universe is to generate consciousness. The Universe spawns consciousness through living beings. The human species may be paltry, but the existence of the spirit in an organism on a planet in the Universe, able to perceive its complexity and its harmony, and to reflect on its meaning, is probably of a fundamental meaning. Man seems to be an important step, but we can however ask ourselves two questions:
- Does he occupy a privileged, even unique place, as the holder of a consciousness capable of thinking about his place in this Universe?
- Do all men wish to access this cosmic consciousness?

Answering the first question comes, as we have seen, from astrophysics and therefore from science. Existential analysis can inspire some reflections in response to the second.

It describes us as ideal human beings where the spirit dominates and uses the instincts, the emotions, the reasonings without stifling them, to put their energy at the service of a noetic dimension. However, we can see that the dimension of the spirit does not concern all of society, for the good reason that a part of it denies it. This dimension began to be rejected, in France, from the Age of Enlightenment, by part of the educated class, seduced by scientific materialism. It was then joined on this point by the anticlerical progressives. As a result, invoking it is interpreted as a sign of reaction or superstition, of political, social or cultural or even intellectual regression. **The historic confrontation between materialism and spiritualism was clearly won, in the 20th century, by materialism.**

Dualism, which affirms the coexistence of spirit and matter, is therefore difficult to assume for anyone who wants to play a public role, since it is considered at the same time, irrational and anti-republican. The reference to the spirit has become a social and political marker, to such an extent that at the beginning of the 21st century, the strong reappearance of a certain radical religiosity, perhaps in reaction to this triumph of materialism, is a social and political phenomenon.

Yet Frankl emphasizes that "Man needs a system of orientation or an object of abandonment; religion or philosophy", and that therefore the question about meaning leads sooner or later to the question about God. Psychology is not there to give answers to the religious questions of men. But a psychology which does not consider the religious dimension of man does not cover its entire domain. Existential analysis has no relationship with religions and does not pronounce on their doctrine, but is interested in religiosity. It recognizes that religion can determine the lives of men, and considers its effect on people's psyches. **In this**

context, the concept of spirituality suffers all the more from being reduced to religion.

Existential analysis conceives the phenomenon of belief not as belief in a God but as belief in an all-encompassing meaning. Frankl's approach to aspiration joins, in a certain way, the reflections of the physicists Penrose or Espagnat, themselves inspired by the cosmic mysticism that some respected thinkers such as Spinoza, Kant and Einstein had already developed. This third way makes it possible to overcome the opposition between religious spiritualists and atheistic materialists, passing from the question of the existence of God to that of the meaning to be found in the existence of the Universe, of life and of consciousness.

We are in fact inevitably confronted with the relationship between "the meaning of one's life" which includes for oneself, projects to be accomplished and a purpose, and "the meaning of life", which is a coherence between human life and existence of the world, which can be called "cosmic meaning". It would then be a spirituality inspired by a Universe carrying a project, even the origin of this project, and which some can possibly qualify as divine instance.

In other words, can we re-enchant Man without re-enchanting the world? It seems possible to find personal meaning in one's life, without resorting to a cosmic system. This is at least the position of Sartre, Camus, and many others, for whom the world only has the meaning that we attribute to it. It has none in itself. The Universe is absurd.

However, is it so easy for a human being who really aspires to discover a purpose, to think of himself in a world that has none? "This finality that we create does not effectively relieve us of our discomfort if we continue to remember that we are the instigator... It is much more comforting to believe that the meaning is out there and that we have only discovered it", writes Irving Yalom, who however does not believe in a superior or cosmic meaning. He also goes on, surprisingly, quoting Frankl: "believing in the meaning of one's life, without believing in the meaning of life, is like wanting to climb a fakir's rope, hanging from nowhere". **It seems established that feeling connected to something bigger, which moreover has meaning, generates a personal feeling of meaning.**

The same Irving Yalom, professor of psychiatry, describes what makes a psychological theory legitimate. He emphasizes how much concepts such as the superego, the id, the ego, the archetypes, the idealized self and the real self, self-esteem, the parent, child and adult states, are fictions, intellectual creations that are not worth only by their explanatory power, and by their capacity to mobilize the will to live: "these explanatory systems operate because they confer a feeling of personal mastery and awaken the weakened will".

Yalom, who like all psychologists aspires for his field to be considered a science, only applies to psychology a definition of scientific theories, which are intellectual creations that are only worth their power to explain and predict phenomena. Thus, a theory supplants the previous one when it better explains the observed events. Einstein's cosmology superseded the previous ones of Newton and Ptolemy for this reason.

Certain psychological interpretations are therefore superior to others, not because they appear deeper, but because they are more effective in their project of reassuring and mobilizing the will to live. The reasoning by efficiency is therefore applicable, and he applies it himself, to the "meaning of his life" in existential psychology. Advocating to find meaning in one's life generates a greater will to live.

Logically, evoking a "meaning of life" to allow the patient to adhere more easily to the need to inscribe "the meaning of his life" therein, would rationally complete the existential theory. But until now, most psychologists, like Irving Yalom, refuse to do so. Why is the mental construction "meaning of one's life" acceptable because it is effective and "meaning of Life" is not effective if it makes the previous one more credible?

It is in fact, because it goes against another mental construction, stemming from the queen of sciences, physics, which until now defined the world as exclusively material and deterministic, devoid of mind and meaning. However, as we have seen, more and more scientists think that the universe, like life, seems to be the result of a project that provides meaning rather than the fruit of absurd chance, the same way many mathematicians think they are discovering mathematics rather than creating them from scratch.

Thus, until now, psychology was not authorized to invoke a cosmic meaning of life or inscribe the earthly meaning of our life, under penalty of not being qualified as science by the global explanatory theory of physics. From now on, then physics, updated, can tolerate, even arouse this hypothesis. **It would therefore be rational and scientific for psychology to induce this hypothesis in its general construction since it provides more efficiency in its raison d'être, which is to develop the will to live.**

In the same way, one can think that religions were mental constructions which were only worth by their capacity to reassure men and to channel their need for transcendence, to tend towards goal higher than themselves. According to the level of knowledge of each era, the effectiveness of such and such a religion proved to be more or less relevant, explaining the evolution of animisms towards polytheisms, and then monotheisms. Today's knowledge expresses the need for consciousness, and admits the limits of a phenomenal reality, and reveals a total reality in itself, much larger and more complex, independent and veiled. They thus provide a growing place for this cosmic spirituality, convinced that the meaning of each life is to be part of the meaning of life and of the Universe.

To think that the Universe exists to generate a consciousness of which we seem to be the only current holders in our visible horizon is dizzying, but to see that we do not exercise that consciousness at the level that can justify our status of a conscious humanity, is distressing. The unconscious personality remains primitive, wild, irrational. If it is not integrated into consciousness through deep psychological work, it can express itself at any time in a violent, barbaric way, both individually and collectively.

The practice of a life integrating a dimension of the spirit, the implementation of the process of individuation, already constituted for Jung the only answers adapted to the evil which corrodes the human being and the world of today.

Changing this world which oscillates between individualistic materialism and religious dogmatism will pass through a general elevation of individual consciousness. Viktor Frankl 's existential analysis is intended to contribute to this.

CONCLUSION

Our civilization has not liberated man as it claims to do. On the contrary, it seems to have partially alienated him by egocentric individualism, consumer vanity, infantilism, even obscurantism.

The American psychologist David Barlow points out how the proportion of people who suffer from depressive disorders has increased significantly over the past half century. Among the plausible explanations, he cites the exacerbation of individualism, the inflation of demands for personal well-being and the decline of the feeling of belonging to greater than oneself, the Fatherland, the Church, the family. This result would come from the voluntary rejection of any reference to a superior entity.

And it actually appears that our society is all the more individualistic, egocentric, consumerist, materialistic, as the educational, social, religious, family structures of yesteryear have weakened, even collapsed. These values previously guided us towards a life oriented on something other than ourselves. Even if it is true that where beliefs have been preserved, the fulfillment of individuals is not always proven, we might have been too proud in our desire for emancipation.

We are weakened by the loss of our ancestral, moral, psychic landmarks, which carried, without our being aware of it, the deep values that were necessary for us. These are no longer sufficiently promoted by our current culture. We need to find them, to reinvent them. **Humanism, long considered as an immense progress and always re-invoked, turns out to be misguided when the man it is supposed to have liberated is nothing more than an egocentric hedonistic individualist.**

Between the process of individualization, stemming from the development of the ego, and that of individuation, stemming from the connection to the self, which Jung clearly distinguished, the first was privileged, creating a society of individualistic people. We now need to reassess the fully individuated being who knows on the contrary the importance of his link to what surrounds him, living beings, nature, the spirit.

Frankl's approach, which recommends seeking this attraction towards something that is beyond us, then reveals all its relevance in the face of this observation. **Existential analysis proposes to the individual to behave as a human being, towards himself and towards others. A human being is one who uses his moral conscience, wants to be free in his will and feels responsible for what he does and what he does not do.** He is the one who aspires to be a contributor to the community of human beings, not a parasite, and a fortiori not a predator.

Viktor Frankl, by asking the question of meaning in psychology, proposed a vision of man endowed with a non-material dimension of the spirit, justifying this need for meaning in his life. This is all the more tangible as it can be part of the conviction of a higher meaning, an ultimate meaning. In all the materialist circles of his time, this position could only be rejected, because contrary to the teachings of science refuting religion and the supernatural.

From now on, science, enriched by many works in all fields, admits on the one hand that matter does not explain all reality, and on the other hand that the Universe seems to carry not only a life project but also a consciousness project, and therefore a supernatural project. Faced with the knowledge of his time, Frankl judged the ultimate meaning as not a matter of thought, but of belief. Today's science makes it possible to reconsider this position. The ultimate meaning is indeed a subject of thought and scientific thought. This ultimate meaning is no longer supernatural, since it comes from Nature. Rationalist thinking is no longer rational. Spirituality must be based on science and no longer exclusively depend on religious dogma. **Science had been created by detaching itself from animism. It must henceforth free itself from materialism.**

Our understanding of the world has evolved considerably over the past century. The physical theory of the Big Bang, with its equations, is today accepted as the most credible for describing the past and the starting point of the cosmos. However, we still do not know where this energy, these forces, these laws, these equations come from, which allowed this starting point to generate in 13.8 billion years, the Universe, and life, and consciousness. But we note that these laws of nature which have allowed this deployment of the World since its creation are part of a pattern determined from the outset.

Man has been able to verify this implacable determinism, which the theory of chaos and quantum physics do not contradict, which show us, not an indeterminism of the future but an indeterminability, that is to say a determinism that we do not know how to measure and predict precisely.

At the same time as the human intelligence noted the realization of this project, the conscience which had emerged from this intelligence proved its aptitude to free itself from this determination hitherto inexorable. Indeed, what comes under the action of a being endowed with a conscience, and therefore with a freedom of will and action, cannot be predetermined, by definition. The exercise of consciousness is an element of indetermination. **That is to say that determinism stops only where the free decision of the exercise of conscience begins.**

In fact, this human consciousness seems to be, in our knowable horizon, the first and only factor of indetermination of the Universe. It alone allows man to decide otherwise than what his deep organic nature would like, therefore what the initial project of the Universe had a priori traced, had predetermined for him. It is possible, even probable, that elsewhere other consciousnesses have already put an end to this total initial determination. We can make the fantastic hypothesis of an initial project of The Universe which foresees, beyond a strictly planned unfolding, a moment when a capacity for decision, creation, independent of the Universe itself must emerge. This moment is the emergence of consciousness from life, then a capacity of this consciousness to influence part of reality for the future.

Science depicts a Universe for us as a project in progress and existential analysis describes a human being who, thanks to his conscience and his spiritual dimension, is free to join it or not, and who is able to modify, no doubt in an infinitesimal way, the course initially determined.

If the appearance of consciousness is inscribed in the project of the Universe, then it has no reason to concern only the human species. It has manifested itself in man for about 100,000 years. Other forms of terrestrial life, the most advanced animals, will also be endowed with a conscience and a spiritual dimension in 100,000 years if they are given time. This perspective should lead us to reconsider all life, and in particular the most evolved, as capable of being, one day in the future, endowed with consciousness and spirit. Our behavior towards them is to be reassessed in its totality. Man is not a separate being. It is now a question of rethinking this concept of humanism, this ideology which recognizes only man, and enrich it with something which is interested in all beings of conscience, current and future.

As for the thesis of a conscious life existing elsewhere, on myriads of other planets, it is now considered very probable by many scientists, so common are the conditions that allowed ours in the Universe. The real question on this subject, therefore, does not concern the existence of extraterrestrial life, but the absence of manifestations of it. The Nobel Prize in Physics Fermi had tried to sum up in a famous paradox the possible reasons for this absence. One of the hypotheses, logical and terrifying, is that these lives self-destruct or regress considerably at a certain stage of development, which prevents any communication over long distances.

We can see that most past human civilizations collapsed at some point in their development for various causes, famines, wars, epidemics, climatic conditions... which they were unable to overcome. Until now, civilizations that have disappeared have been replaced by others. From now on, civilization is planetary, and many are those who think that it is in danger.

However, life in multiple planets seems to us to have been registered from the beginning in the initial project of the Universe, according to an inescapable determinism resulting from the physical laws of the origins. This project of the Universe, determined although partially unpredictable, written from time immemorial, since time begins for us at the birth of our Universe, would therefore organize a self-destruction or a regression of forms of life when they reach a certain level of technicality. Or at least those who would not have taken advantage of their conscience to tear themselves away from the determinism of preconscious forces, the instincts, the drives, the emotions, the rational intelligence, which brought them to ruin.

This would mean that the acquisition of consciousness, determined from the beginning, is the opportunity given to the civilization that uses it, to escape materially determined destruction. And the absence of signals from other extraterrestrial life would show that this life-saving exercise in consciousness is not common, and therefore not acquired. It is very possible that we are at this stage of decision.

The meaning of the Universe would thus paradoxically be to create in a deterministic way a consciousness that allows life to escape this determinism, by exercising wisely the freedom that this consciousness brings. In this case, Frankl had grasped part of the role of conscience, that of giving meaning to one's life, to everyone's life. But in fact, consciousness would also make it possible to give meaning to Life, in the Universe.

If man has a place in the project of the Universe, he will only occupy it by rising to the height of this destiny. We know that we are not creating the order of the Universe, but we are growing in our awareness of that order which is already there. We were able to perceive that nothing is definitively acquired in this ascent and that the regressions are too frequent and always hopeless. Our human civilization is fragile, fallible and on the way to self-destruction, for lack of sufficient awareness.

In this obviously intimidating context, existential analysis constitutes above all a reflection on the human condition, underlining the urgency of distinguishing between determinism and free will, matter and spirit. Frankl was convinced that the specificity of man lay in his dimension of the spirit, which was expressed by this consciousness bearing freedom, responsibility, and the need for meaning. If Frankl's thought seduces today by responding to a certain growing demand for the quest for meaning or spirituality, it also repels by its other dimension, yet inseparable, of responsibility and freedom. This thought, however, in no way disdains those who suffer, the victims of injustice or of a miserable fate, and the material help that can and should be given to them, but it considers that this help will always be incomplete as long as it will not have aroused the strength and the will to place life in a direction that goes beyond it. **It is the essence of Existential Analysis to allow man to free himself from the shackles of his past, to free himself from the weight of the present, to overcome all that has been described to him as an insurmountable determinism, thanks to this dimension of the spirit that we will have helped him to identify and manifest.**

...

Please, send your comments or corrections to gillet.joel@gmail.com

If you liked this book, please write a review on the site where you purchased it. Thank you, it makes all the difference.

Also from the same author: "Humanizing Business" (logotherapy and management), on Amazon.

Referenced Authors

Alfred **Adler** (1870-1937): Austrian physician, psychotherapist; founder of individual psychology.

David **Barlow:** American psychologist, professor of psychology and psychiatry at Boston University.

Niels **Bohr** (1885-1962): Danish physicist, chemist, philosopher of science; professor at the University of Copenhagen; Nobel prize in physics.

Jean-Pierre **Changeux**: French neurobiologist, geneticist; Professor at the Pasteur Institute and at the College de France.

Remy **Chauvin** (1913-2009); French biologist; professor at the University of Strasbourg.

Nicolas **Copernicus** (1473-1543): Polish astronomer, physician, mathematician; father of the theory of heliocentrism.

Francis **Crick** (1916-2004): British biologist; Nobel Prize in Physiology.

Antonio **Damasio**: Portuguese doctor, neurologist, psychologist; college professor in California.

Anne **Dambricourt** : French paleoanthropologist; research director at the CNRS and the University of Compiègne.

Charles **Darwin** (1809-1882): English naturalist, paleontologist, philosopher, writer; father of the modern theory of evolution.

Paul **Davies**: British physicist, writer; university professor in cosmology.

Christian **de Duve** (1917-2013): Belgian biologist, physician; Nobel Prize in Medicine.

Gilles **Deleuze** (1925-1995): French philosopher; university professor in Paris.

Rene **Descartes** (1596-1650); French mathematician, philosopher, physicist; inspiration of Cartesianism and modern philosophy.

Paul **Dirac** (1902-1984): British mathematician, physicist; Nobel prize in physics.

Albert **Einstein** (1879-1955): German physicist; Nobel prize in physics.

Enrico **Fermi** (1901-1954): Italian-American physicist; Nobel prize in physics.

Cynthia **Fleury**: French philosopher and psychoanalyst; professor at the National Center for Arts and Crafts.

Viktor **Frankl** (1905-1997): Austrian psychiatrist, psychologist, neurologist, philosopher; professor at the University of Vienna; founder of logotherapy/existential analysis.

Sigmund **Freud** (1856-1939): Austrian neurologist, physician; founder of psychoanalysis.

Galileo (1564-1642): Italian physicist, mathematician, astronomer, surveyor; considered the founder of modern physics.

Michael **Gazzaniga**: American psychologist, neurobiologist; university professor in New York.

Kurt **Gödel** (1906-1978): American logician and mathematician; professor at Princeton University.

Yuval Noah **Harari** : Israeli historian; professor at the University of Jerusalem.
Stephen **Hawking** (1942-2018): British physicist, cosmologist; professor at the University of Cambridge.
Martin **Heidegger** (1889-1976): German philosopher; one of the most important philosophers of the 20th century.
Werner **Heisenberg** (1901-1976): German physicist; Nobel Prize in Physics in 1932; one of the founders of quantum mechanics.
Edwin **Hubble** (1889-1953): American astronomer; demonstrated the existence of other galaxies besides the Milky Way.
Edmund **Husserl** (1859-1938): Prussian philosopher, logician; founder of phenomenology.
Karl **Jaspers** (1883-1969): German-Swiss philosopher, psychiatrist; one of the fathers of existentialism.
Carl Gustav **Jung** (1875-1961): Swiss psychiatrist, psychotherapist; founder of analytical psychology.
Immanuel **Kant** (1724-1804): Prussian philosopher; exerted a considerable influence on modern phenomenology and philosophy.
Johannes **Kepler** (1571-1630): German astronomer, mathematician; mathematically defined the trajectories of the planets.
Soren **Kierkegaard** (1813-1855): Danish philosopher, writer, theologian; father of Christian existentialism.
Pierre-Simon de **Laplace** (1749-1827): French mathematician, physicist, astronomer, politician.

Georges **Lemaitre** (1894-1966): Belgian physicist, astronomer, mathematician; professor at the University of Louvain.

Frédéric **Lenoir** : French sociologist, writer, journalist;

Benjamin **Libet** (1916-2007): American neurobiologist, psychologist; university professor in San Francisco.

Karl **Marx** (1818-1883): Prussian philosopher, historian, sociologist, economist, journalist, theoretician of the revolution.

Abraham **Maslow** (1908-1970): American psychologist, sociologist; university professor; father of the humanistic approach in psychology.

Rollo **May** (1909-1994): American existentialist psychologist, writer, psychotherapist; university professor at Harvard and Yale.

Jacques **Monod** (1910-1976): French biologist, biochemist; Nobel Prize in Medicine.

Isaac **Newton** (1643-1727): English mathematician, physicist, astronomer, philosopher; professor at the University of Cambridge.

Blaise **Pascal** (1623-1662): French mathematician, physicist, philosopher, theologian.

Roger **Penrose**: British mathematician, cosmologist, philosopher of science; Nobel Prize in Physics 2020; professor at Oxford University.

steven **pinker** : Canadian psychologist; philosopher, anthropologist; university professor at Harvard and Stanford.

Max **Planck** (1858-1947): German physicist; Nobel price physics.

Otto **Rank** (1884-1939): Austrian psychologist and psychoanalyst.

Carl **Rogers** (1902-1987): American humanist psychologist; founder of the psychotherapy called "person-centered approach".

Ernest **Rutherford** (1871-1937): New Zealand physicist, chemist; father of nuclear physics. Nobel Prize in Chemistry.

Jean-Paul **Sartre** (1905-1980): French existentialist philosopher, writer; Nobel Prize in Literature, which he refused.

Max **Scheler** (1874-1928): German sociologist, philosopher, anthropologist; university professor in Cologne.

Erwin **Schrödinger** (1887-1961): Austrian physicist; Nobel price physics.

George **Smooth:** American astrophysicist, cosmologist; Nobel price physics.

Trinh Xuan **Thuan** : Vietnamese-American astrophysicist; professor at the University of Virginia.

John **Wheeler** (1911-2008): American physicist; professor at Princeton University.

Irvin **Yalom**: American psychiatrist, existentialist psychotherapist; professor at Stanford University.

Thomas **Young** (1773-1829): English physicist, physician; famous for his optical experiments.

Bibliographic references

Bitbol Michael, 2000. *Physique et philosophie de l'esprit*, Paris, Flammarion.
Damasio Antonio, 2017 . *L'Ordre étrange des choses*, Paris, Odile Jacob.
D'Espagnat Bernard, 2002. *Traité de physique et de philosophie*, Paris, Fayard. 2015. *A la recherche du réel*, Paris, EHKO. 2014. *Le monde quantique*, Paris, Editions Matériologiques.
Frankl Viktor, 2017. *Retrouver le sens de la vie*, Paris, Interéditions. 2019. *Nos raisons de vivre*, Paris, Interéditions.
Fleury Cynthia, 2020. *Ci-gît l'Amer*, Paris, Gallimard.
Harari Yuval Noah, 2015. *Sapiens, une brève histoire de l'humanité,* Paris, Albin Michel.
Hawking Stephen, 1989. *Une brève histoire du temps*, Paris, Flammarion.   2011. *Y a-t-il un grand architecte dans l'Univers ?* Paris, Odile Jacob.

Kühn Rolf, 2015. *Logothérapie et phénoménologie*, Paris, L'Harmattan.

Laplane Dominique, 2005. *Penser, c'est-à-dire ?* Paris, Armand Colin.

Lenoir Frédéric, 2014. *Le Christ philosophe*, Paris, Plon.

Le Vaou Pascal, 2006. *Une psychothérapie existentielle. La logothérapie de Viktor Frankl*, Paris, L'Harmattan.

Meyer Catherine, 2008. *Les nouveaux psys.* Paris, Les Arènes.

Penrose Roger, 2019. *Les deux infinis et l'esprit humain,* Paris, Champs.

Pinker Steven, 2018. *Le triomphe des Lumières*, Paris, Actes Sud. 2017. *La part d'ange en nous,* Paris, Les Arènes.

Reeves Hubert, 2017. *Le banc du temps qui passe*, Paris, Le Seuil.

Sartre Jean-Paul, 2001. *L'existentialisme est un humanisme*, Paris, Gallimard.

Staune Jean, 2007. *Notre existence a-t-elle un sens ?* Paris, La Renaissance.

Yalom Irvin, 2010. *Thérapie existentielle,* Galaade.

Zwirn Hervé, 2000. *Les limites de la connaissance*, Paris, Odile Jacob.

www.ingramcontent.com/pod-product-compliance
Lightning Source LLC
Chambersburg PA
CBHW050004230526
45465CB00003BB/1256